喚醒你的英文語感！

Get a Feel for English !

喚醒你的英文語感 ！

Get a Feel for English !

5句話搞定商務Eail!

100則電郵範本，隨查即用

作者 / Shawn M. Clankie、小林敏彥　總編審 / 王復國

mail

SAVE　SEND

E-mail　回覆　刪除　回覆　全部回覆　轉寄　　附件　　移動　垃圾　　本週　分類　追查　篩選　　　搜尋連絡人　　　　　　登出/接收

▶ **Subject: Payment sent**

Ryan,

Just following up to let you know that I deposited the money for the office furniture into your corporate account this afternoon. It should already appear in your account. If there are any problems, let me know. It was a pleasure doing business with you.

Yukiko Sanada
Moana Surfboards

SEND出你的競爭力

10大主題＋超過400個好用句

- ☑ 預約會面
- ☑ 合約訂單
- ☑ 公告祝賀
- ☑ 商務往來
- ☑ 客訴處理
- ☑ 問候祝福

貝塔語言出版
Beta Multimedia Publishing

IRT 語言測驗中心
Language Testing Center

序

　　本書集結了 100 篇以單一段落（原則上在 5 句內）構成且簡潔易懂之英文商務書信範例，是一本英文 email 寫作手冊。在這些切中要點的範例中，除了介紹可直接照抄運用的基本文句和相關句型外，另外更進一步以 Q&A 形式解說與英文商務電郵有關的表達方式及異國文化資訊。

　　筆者兩人一起在商學院（提供專業人士取得日本國內 MBA 學位之課程的研究所）負責教授商務英語（Team Teaching‧協同教學）。由於每天和日本商務人士密切相處與上課，故能充分掌握其需求，經由分析企業內部及企業彼此間實際往來的電子郵件，才完成本書所收錄的眾多範例。而除了商務需求外，也包含其他諸如聖誕節祝賀與婚喪喜慶……等，對個人亦極具參考價值的例子。書中介紹了許多英文 email 常見例句、詞彙，不僅能幫助讀者於短時間內讀懂收到的英文信，更是一本在繁忙生活中無須多花時間，即可提升英文寫作力的【應用型英文 email 範例集】。

　　我們的生活因網路普及而徹底改變。透過電子郵件的運用，讓人們能輕易地相互連結、密切溝通，更可輕鬆地取得聯繫。尤其在商業領域，email 早已成為基本通訊方式，利用其便利性，無論公司內外事務都可搞定。而一天 24 小時隨時收發信件也成了基本商務型態。可迅速傳達資訊，還能節約用紙的電子郵件，對國際化的商務人士而言，確實是不可或缺的溝通工具。

　　本書不僅可當成範例集運用，更建議讀者參考書中範例，將今後可能用到的信件內容事先建立為範本儲存起來。商務 email 只要事先準備好範本，再以替換詞彙的方式分別反覆利用，即可有效提升生產力。我們誠心希望本書對於提升各位的英語溝通力能有所助益。

　　最後，同樣要感謝日本株式會社語研公司奧村民夫先生對本書編輯工作的大力支援。我們深信本書是融入了大量企業觀點的實用集，在此向他再度致上最高謝意。

北方有一顆星，雖小，其光輝卻如此明亮

國立大學法人小樽商科大學語言中心副教授　**Shawn M. Clankie**
國立大學法人小樽商科大學商學院研究所教授　小林敏彥

CONTENTS 目錄

序 3

1 預約會面 Business Appointments

1. 提出會面邀約 Requesting an Appointment ················· 12

2. 回覆會面邀約 Responding to a Request for a Business Appointment ······ 14

3. 約定時間、地點 Scheduling the Meeting Date and Place ················· 16

4. 取消會面 Cancelling the Appointment ················· 18

5. 會面後的後續連絡 Following-up after a Meeting ················· 20

2 商務往來 Doing Business

6. 提案 Suggesting a New Idea ················· 24

7. 回應提案 Responding to a Suggestion ················· 26

8. 突顯商品的賣點 Highlighting a Product's Selling Points ················· 28

9. 針對商品提問 Inquiring about a Product ················· 30

10. 推薦企業／人才 Recommending a Company / Person ················· 32

11. 索取文件或資訊 Requesting Documents or Information ················· 34

12. 確認進度 Checking the Progress of Work ················· 36

13. 確認庫存 Checking Stock ················· 38

14. 聯繫某個部門或團隊 Contacting a Department or Team ················· 40

15. 確認已接收 Acknowledging Receipt ················· 42

3 合約‧訂單 Contract / Order

16. 請求報價 Requesting an Estimate ················· 46

17. 議價 Negotiating a Discount ················· 48

18. 提出折衷價格 Offering a Compromise Price ········· 50

19. 交涉合約內容 Negotiating a Contract ········· 52

20. 同意合約內容 Agreeing to a Contract ········· 54

21. 修改合約內容 Modifying a Contract ········· 56

22. 提出折衷方案 Offering a Compromise on a Contract ········· 58

23. 解約通知 Notifying Annulment of a Contract ········· 60

24. 寄送合約書 Sending a Contract ········· 62

25. 訂購商品（下單）Ordering a Product ········· 64

26. 修改訂單（改單）Changing an Order ········· 66

27. 確認訂單 Confirming an Order ········· 68

28. 要求出貨 Requesting Shipment ········· 70

29. 出貨通知 Notifying Delivery ········· 72

4 付款 Payment

30. 請求付款 Requesting Payment ········· 76

31. 確認付款方式 Asking about the Method of Payment ········· 78

32. 寄送發票 Sending an Invoice ········· 80

33. 請求重開帳單 Requesting a Reissued Bill ········· 82

34. 已付款通知 Notifying of a Payment Made ········· 84

35. 確認已收款 Acknowledging Receipt of Payment ········· 86

36. 要求支付拖欠帳款 Requesting Payment for a Delinquent Account ···· 88

37. 請求延後付款 Asking for Deferment ········· 90

5 客訴處理・道歉 Trouble Management / Apology

38. 投訴產品問題 Complaining about a Product ········· 94

39. 針對產品瑕疵道歉 Apologizing for a Defective Product ········· 96

40. 針對缺貨問題道歉 Apologizing for Something out of Stock ········· 98

41. 投訴貨品運送延誤 Complaining about a Delivery Delay ·········· 100

42. 針對貨品運送延誤道歉 Apologizing for a Delivery Delay ·········· 102

43. 投訴客服不周 Complaining about Customer Service ················ 104

44. 針對客服問題道歉 Apologizing for Customer Service ················ 106

✉ **6** 公告‧祝賀 **Business Announcements / Congratulations**

45. 漲價通知 Announcing a Price Increase ························· 110

46. 新產品發表 Announcing a New Product ························· 112

47. 祝賀新事業 Congratulating on a New Business ·················· 114

48. 退休／調職通知 Announcing Someone's Retirement / Transfer ········· 116

49. 職務異動通知 Announcing a Change in Job Assignment ··········· 118

50. 公告新地址 Announcing a New Address ······················· 120

51. 休假（暫停營業、暫時離開）通知 Announcing a Holiday Break ······· 122

52. 新任職者到任通知 Announcing a New Hire ····················· 124

53. 員工訃聞通知 Breaking News of an Employee's Death ············· 126

54. 回應員工訃聞 Responding to News of an Employee's Death ········· 128

55. 調動、辭職及退休通知

Notifying a Transfer, Resignation and Retirement ·············· 130

56. 回應調動、辭職及退休通知

Responding to a Notification of a Transfer, Resignation and Retirement ······· 132

57. 歡送會通知 Notifying a Farewell Party ························· 134

58. 請病假 Taking a Sick Day ····························· 136

✉ **7** 日常業務 **Daily Work**

59. 會議通知 Announcing a Meeting ························· 140

60. 寄送會議紀錄與報告 Sending Minutes and Reports ················ 142

61. 腦力激盪會議通知 Announcing a Brainstorming Meeting ············· 144

62. 專案參與邀請 Requesting to Participate in a Project ··············· 146

63. 請求提供資訊 Asking for Providing Information ················· 148

64. 回應合作邀請 Responding to a Request for Cooperation ············· 150

65. 請求介紹 Requesting an Introduction ······················· 152

66. 感謝介紹人 Showing Appreciation for an Introduction ·········· 154

67. 宣告公司新規定 Announcing a New Office Regulation ·········· 156

68. 分享資訊 Sharing Information ·········· 158

69. 伺服器維修公告 Announcing Server Maintenance ·········· 160

70. 公告新辦公設備 Announcing New Office Equipment ·········· 162

71. 辦公室裝修通知 Announcing Office Renovation ·········· 164

8 出差 Business Trips

72. 出差指派通知 Ordering Someone to Make a Business Trip ·········· 168

73. 請求安排出差事務 Requesting Travel Arrangements ·········· 170

74. 出差報告 Reporting a Business Trip ·········· 172

75. 出差費用請款 Requesting Reimbursement of Expenses ·········· 174

9 個人問候・祝賀 Personal Greetings / Congratulations

76. 祝賀聖誕節 Christmas Greetings ·········· 178

77. 祝賀新年 New Year Greetings ·········· 180

78. 祝賀情人節 Valentine's Day Greetings ·········· 182

79. 祝賀生日 Celebrating a Birthday ·········· 184

80. 畢業祝賀 Celebrating Graduation ·········· 186

81. 祝賀升遷 Celebrating a Promotion ·········· 188

82. 退休祝賀 Celebrating Someone's Retirement ·········· 190

83. 祝賀喬遷 A Housewarming Message ·········· 192

84. 祝賀訂婚 Celebrating an Engagement ·········· 194

85. 祝賀新婚 Celebrating a Marriage ·········· 196

86. 祝賀生產 Celebrating the Birth of a Child ·········· 198

87. 敬祝早日康復 Get-well Messages ·········· 200

88. 喪事弔唁 Sympathy ·········· 202

89. 寄送賀禮 Sending a Present ·········· 204

90. 收到賀禮後表達謝意 Receiving a Present ·········· 206

10 私人事務通知 Personal Announcements

91. 電話號碼‧地址變更通知

  Announcing a New Phone Number or New Address ·············· 210

92. 搬家通知 Announcing a Move ·············· 212

93. 畢業通知 Announcing One's Graduation ·············· 214

94. 升遷通知 Announcing One's Promotion ·············· 216

95. 新居落成通知 Announcing a New Home ·············· 218

96. 訂婚‧結婚通知 Announcing One's Engagement and Marriage ·············· 220

97. 產子通知 Announcing the Birth of a Child ·············· 222

98. 出院通知 Announcing Getting out of the Hospital ·············· 224

99. 附上照片 Attaching Pictures ·············· 226

100. 對所附照片的回應 Responding to a Message with Pictures ·············· 228

附錄　隨查即用！商英 Email 好用句 400

開頭敬稱 ·············· 232

信尾敬詞 ·············· 232

起頭句 ·············· 233

附件 ·············· 234

維繫關係 ·············· 235

請託 ·············· 235

提案 ·············· 237

提供協助 ·············· 237

會議 ·············· 238

通知 ·············· 240

投訴‧抱怨 ·············· 240

催促 ·············· 242

季節寒暄祝福 ·············· 243

感謝 ·············· 243

道歉 ⋯⋯⋯⋯⋯⋯⋯⋯⋯⋯⋯⋯⋯⋯⋯⋯⋯⋯⋯⋯⋯⋯ 246

祝賀 ⋯⋯⋯⋯⋯⋯⋯⋯⋯⋯⋯⋯⋯⋯⋯⋯⋯⋯⋯⋯⋯⋯ 247

贈禮 ⋯⋯⋯⋯⋯⋯⋯⋯⋯⋯⋯⋯⋯⋯⋯⋯⋯⋯⋯⋯⋯⋯ 249

傷病相關 ⋯⋯⋯⋯⋯⋯⋯⋯⋯⋯⋯⋯⋯⋯⋯⋯⋯⋯⋯⋯ 250

喪事弔唁 ⋯⋯⋯⋯⋯⋯⋯⋯⋯⋯⋯⋯⋯⋯⋯⋯⋯⋯⋯⋯ 250

✉ **本文編號說明**

在每一篇郵件範文中所標示的數字表示該例句為全篇第幾句話,一封成功的商務英文 Email 必須在短短 5 句話之內,將發信者所要表達的重點簡潔地傳達給對方,如此才能在繁忙的工作中事半功倍地完成任務!

1

預約會面
Business Appointments

1　提出會面邀約
Requesting an Appointment

談生意就從預約會面開始。切勿唐突地寄出長篇郵件，而應該先用短短數行的簡潔郵件探探對方意願。另外，務必寫上郵件主旨。

Subject: New development project

Dear Sharon Mayer,

[1]We are planning to start working on a new development project for SmartInfo Co. [2]If it is possible, I would like to see you to hear your opinion on the project. [3]Please let me know when is convenient for you so that I can come to see you at that time. [4]I look forward to hearing from you.

Sincerely,

Tony Kobayashi

翻譯

主旨：新專案開發

Sharon Mayer 小姐，您好：
我們計畫協助 SmartInfo 公司開始進行一項新專案。如果可以的話，我希望能和您見個面，並聽聽您對此專案的意見。請告知您方便的日期、時間，以便我前往拜訪。期待您的回覆。

小林 Tony
敬啓

 重要片語 & 句型

- **We are planning to** *V.* 我們計畫做～。
- **work on ...** 著手進行～
- **if (it is) possible ...** 如果可以的話
- **I would like to see you to hear your opinion on ...**
  我希望能和您見個面，並聽聽您對～的意見。
- **Please let me know when (it) is convenient for you.** 請告知您方便的日期、時間。
- **I look forward to hearing from you.** 期待您的回覆。

佳句便利貼

■ I wonder if it would be convenient for you to see me next week.
不知您下週是否方便與我見個面。

■ Could you spare about 30 minutes with me?
可以給我 30 分鐘左右的時間嗎？

■ Would it be possible to set up an appointment with you on June 28?
是否可與您約在 6 月 28 日？

■ Is it possible for me to see you?
是否能與您見個面？

■ I am prepared to accommodate my schedule to yours.
我可以調整行程以配合您的時間。

■ I would appreciate a brief visit with you tomorrow to introduce our new project. 我希望明日能簡短地拜訪您，並介紹我們的新專案。

■ I would appreciate it if you could give me a call as soon as possible.
如果您能儘早給我通電話，我將非常感激。

FAQ 1　敲定生意的關鍵

Question

What do we do if we do not know who to send our email message to?

如果不知 email 的收件者是誰的話，該怎麼辦？

Answer

We have a set phrase in English that we can use for an opening to a message where we don't know the recipient. It is "To whom it may concern." "To whom it may concern" tells the reader that the message is for anyone that is relevant. We can use it for email messages, such as requesting information from a company, or when we make first contact. The phrase, however, is sometimes viewed as unprofessional, so it is best to try to find the name of a contact person in the company.

對於不知收件者的 email，有種固定的英文開頭寫法可用，那就是「To whom it may concern」（敬啓者）。這個片語能告知讀信者，這封信是寫給所有相關人士看的，故可應用於向公司索取資料或首度接觸的 email 中。但是這樣的開頭有時會被認為不夠專業，所以最好還是試著找出該公司聯繫人姓名比較理想。

2 回覆會面邀約
Responding to a Request for a Business Appointment

收到會面邀約時，務必確認行程並及早回信。若無法配合對方所指定的行程，就自行提出方便的日期、時間。

Subject: Re: Appointment request

Mr. Adams,

[1]My name is Kumiko Tanimoto and I'm responsible for sales at Muroran Steel. [2]Thank you for your request for an appointment. [3]We will be available to meet with you on July 10th at 2 p.m. [4]If this is acceptable, please let me know and we will finalize the details. [5]Thank you.

Best Regards,

Kumiko Tanimoto

翻譯

主旨：回覆：預約會面

Adams 先生，您好：

我是負責 Muroran 鋼鐵公司業務工作的谷本久美子。感謝您來信預約會面。我們可以在 7 月 10 日下午 2 點跟您見面。如果沒問題，就請通知我一聲，以便我們敲定細節。謝謝。

祝好
谷本久美子

 重要片語 & 句型

■ **I'm responsible for** ... 我負責～。
■ **Thank you for your request for an appointment.** 感謝您來信預約會面。
■ **We will be available to meet with you on** ... 我們可以在～跟您見面。
■ **on** *A* **at** *B* 於 A 日期 B 時間
■ **If this is acceptable, please let me know.** 如果沒問題，就請通知我一聲。
■ **finalize the details** 敲定細節

佳句便利貼

■ Let me check his/her schedule and get back to you.
讓我先確認一下他／她的行程，再回覆您。

■ We could schedule you at 1 p.m. on Monday.
我們可以將您安排在週一下午 1 點。

■ Paul Jones can see you next week.
Paul Jones 下週可與您會面。

■ What is this regarding?
會面目的為何？

■ Ms. Crosby is booked solid until May.
Crosby 小姐直到 5 月為止都行程滿檔。

■ I'm sorry but Mr. Ali is only available at that time.
很抱歉，Ali 先生只有那個時候有空。

■ I'm afraid Mr. Palani is not available.
恐怕 Palani 先生無法與您會面。

FAQ 2　敲定生意的關鍵

Question
Do we always have to start an email message with "Dear"?

電子郵件一定要以「Dear」開頭嗎？

Answer
"Dear" in the opening of an email is not always necessary (Dear Mr. Reynolds, Dear Ms. Adams) in business email messages. It is natural and appropriate simply to use the person's name (Mr. Reynolds, Ms. Adams). First names should only be used when there is a long-established relationship with the person you are writing to, or if you are co-workers.

撰寫英文商務 email 時，並不一定要用「Dear」起頭（例如 Dear Mr. Reynolds、Dear Ms. Adams）。僅使用對方姓名（如 Mr. Reynolds、Ms. Adams）也是很自然並且合宜的。而只有在已與對方建立長久合作關係，或為同事身分時，才可直呼其名（first name）。

約定時間、地點
Scheduling the Meeting Date and Place

具體提出會面時間與地點。如果要在咖啡廳或餐廳會面，就該以對方方便為主，指定清楚易懂的地點。

Subject: Meeting

Ms. Sanchez,

[1]Thank you for offering to meet so that we might discuss our mutual business interests. [2]Might I suggest that we meet over lunch at La Piazza on 12th Street? [3]How does 1 p.m. sound? [4]If this works for you, please let me know and I will go ahead and make the reservation. [5]Looking forward to seeing you.

Sincerely,

Takako Endo

翻譯

主旨：會議

Sanchez 小姐，您好：
感謝您願意見面以便討論我們的互利業務。我可否建議約在第 12 街的 La Piazza 餐廳，邊吃午飯邊聊？下午 1 點如何？如果沒問題，就請通知我一聲，我會去訂位。期待與您見面。

遠藤貴子
敬啟

 重要片語 & 句型

■ **Thank you for offering to meet so that ...** 感謝您願意見面以便～。
■ **Might I suggest that ...?** 我可否建議～？（比 May I suggest...? 更有禮貌）
■ **How does ... sound?** ～如何？
■ **If this works for you, please let me know.** 如果沒問題，就請通知我一聲。
■ **I will go ahead and V.** 我就會繼續進行～。

佳句便利貼

■ This calls for a meeting.
這需要開會討論。

■ We'd like to have a meeting on Wednesday at 1 in the conference room.
我們想在週三下午 1 點於會議室開個會。（Note：依序寫出日期、時間、地點）

■ The meeting will be held at the convention center.
會議將在會議中心舉行。

■ We'll be meeting in Meeting Room 4.
我們將在第 4 會議室見面。

■ Let's set up a working lunch.
讓我們訂個午餐會吧。

■ Let's discuss this over dinner.
讓我們邊吃晚飯邊談吧。

■ There is a mandatory meeting at 5.
5 點有個強制參加的會議。

 **FAQ 3** 敲定生意的關鍵

Question
How should we close a business email message? 商務 email 該如何結尾？

Answer

Business email messages generally end with a closing, followed by the person's name and the person's position if it is not known to the recipient of the message. "Sincerely" and "Best regards" are the two standard closings for business email messages.

商務 email 通常會在最後寫上結尾詞，再接著寫人名、職稱（如果對方不知道的話）。而「Sincerely」和「Best Regards」是兩種商務 email 的標準結尾詞。

如果要取消約定，務必提早通知對方。而依據取消原因不同，有時最好能進一步以電話補充說明。

Subject: Re: October 4th meeting

Mr. Jameson,

[1]I am sorry to inform you that I must cancel our meeting scheduled for October 4th. [2]Unfortunately, my plans to be in your area at that time have fallen through. [3]If possible, on my next trip to Boston I would like to try to reschedule. [4]I apologize for this inconvenience and hope we can meet on my next trip.

Sincerely,

Gen Mastumoto
President, Matsumoto Industries

翻譯

主旨：回覆：10 月 4 日的會議

Jameson 先生，您好：

很遺憾要通知您，我必須取消 10 月 4 日的會面約定。很可惜，我原定當天要前往貴公司附近拜訪的行程泡湯了。如果可以的話，我下次去波士頓時希望能再跟您約一次。很抱歉造成您的不便，希望下次出差能有機會與您見面。

松本企業 總裁
松本源
敬上

 重要片語 & 句型

■ **I am sorry to inform you that ...** 很遺憾要通知您～。
■ **cancel** *one's* **meeting scheduled for ...** 取消～的會面約定
■ **fall through** 失敗；告吹
■ **reschedule** 重新安排
18 ■ **I apologize for this inconvenience.** 很抱歉造成您的困擾。

佳句便利貼

■ I can't make it to the meeting on Friday.
我沒辦法參加週五的會議。

■ I have some business to attend to.
我有些事要處理。

■ Would it be possible to reschedule?
可以重新安排一個時間嗎？

■ Could we meet on a different day?
我們可以約其他日子見面嗎？

■ Is there an alternate day we can get together?
我們可以約其他日子聚聚嗎？

■ I apologize for cancelling at the last minute.
我很抱歉在最後一刻取消約定。

■ I have no choice but to cancel the meeting.
我不得不取消會議。

 **FAQ 4** 敲定生意的關鍵

Question
What if I must cancel at the last minute?　　　如果必須在最後一刻取消約定，該怎麼辦？

Answer

If you must cancel at the last minute, then a telephone call is preferred to an email message. It is seen as more professional and, at the last minute, it is possible that the recipient may not see your message in time.

如果你必須在最後一刻取消約定，那麼打電話取消會比用 email 好。打電話通知會顯得比較專業，而且都到了最後一刻，收件人很可能無法即時看到你的 email。

與對方會面後，務必發一封 email 表達感謝之意。事實上不僅限於會面之後，在派對或餐會後都該寄送 email 或卡片，才符合商務禮儀。

Subject: Thank you

Dear Ms. Stevens,

[1]Thank you once again for taking the time to meet with me last Friday. [2]It was a pleasure speaking with you about our products and I hope that we will have the opportunity to do business together now and in the future. [3]If I could offer any further information, please do not hesitate to contact me.

Sincerely,

Ken-ichi Kanazawa

翻譯

主旨：感謝您

Stevens 小姐，您好：
再次感謝您上週五撥冗與我會面。很榮幸能與您談談敝公司產品，誠摯希望從而後能有機會與您合作。如果您需要任何進一步資訊，請不吝與我聯繫。

金澤健一
敬上

 重要片語 & 句型

- **Thank you once again for taking the time to meet with me.** 感謝您撥冗與我會面。
- **It was a pleasure speaking with you.** 很榮幸能與您談話。
- **I hope that we will have the opportunity to do business together.**
  希望有機會能與您合作。
- **now and in the future** 從今而後
- **If I could offer any further information, please do not hesitate to contact me.**
  如果您需要任何進一步資訊，請不吝與我聯繫。

佳句便利貼

■ It was an honor to meet you.
十分榮幸能與您見面。

■ I really enjoyed our conversation.
和您聊得真的很開心。

■ I'm grateful for the opportunity to meet with you.
我很感激有機會能與您見面。

■ If I can be of any more help, please let me know.
如果還有什麼能幫得上忙的，就請通知我一聲。

■ Please contact me, if you need any further assistance.
如果您需要進一步協助，請與我聯繫。

■ Thank you for your kindness and hospitality.
感謝您的和善與接待。

■ Your kindness is greatly appreciated.
非常感謝您的和善。

 **FAQ 5** 敲定生意的關鍵

Question

How soon should I send a follow-up message after a business meeting?

應該在商務會面後多久，發送後續的 email？

Answer

Following up after an important meeting is always important. But you don't have to email immediately, especially if you are on a business trip. Sending a follow-up message the next day or even the next week (if you are travelling) is fine. The follow-up helps keep you in the minds of the people you met with in the meeting.

重要商務會面之後的後續連絡非常重要，但是你並不需立刻寄送 email，尤其是在出差的情況下。一般來說，隔天或隔週（如果你在旅途上）再發信就行了。而這種後續的 email，可讓會面對象對你留下更深刻的印象。

Q&A

2

商務往來
Doing Business

提案
Suggesting a New Idea

要提出新提案的時候，應先做簡介，然後觀察對方是否有興趣。同時準備好詳細的介紹文章與附件資料，以便收到積極回應時能立即傳給對方。

Subject: Flex time proposal

Mr. Tilley,

[1]I have this idea I've been working on. [2]I'd like to run it by you. [3]How about if we offer flex time to all employees? [4]This would allow employees a more flexible work environment and would increase productivity. [5]If you need further details, please let me know and we can get together and talk about it.

Thanks,

Sho Ishihara

翻譯

主旨：彈性工時提議

Tilley 先生，您好：

我一直有個想法，希望能說給您聽聽，看看您有什麼意見。您覺得對所有員工實施彈性工時制如何？這樣做能讓員工享有更具彈性的工作環境，也能增加生產力。如果您需要更多詳細資料，請通知我一聲，我們可以見面討論。

謝謝。
石原翔

 重要片語 & 句型

■ **I have this idea I've been working on.** 我一直有個想法。
■ **I'd like to run it by you.** 希望能說給您聽聽，看看您有什麼意見。
■ **How about if ...?** ～的話您覺得如何？
■ **If you need further details, please let me know.**
　如果您需要更多詳細資料，請通知我一聲。
■ **We can get together and talk about it.** 我們可以見面討論。

佳句便利貼

- **Here's an idea!**
  我有個好點子！

- **I have a great idea.**
  我有個很棒的點子。

- **Allow me to make a suggestion.**
  容我提出一個建議。

- **I'd like to make a proposal.**
  我想提個案。

- **What if we do this instead?**
  如果改成這樣做呢？

- **Why don't we consider this idea?**
  我們何不考慮採用這個構想？

- **Wouldn't it make sense to try something new?**
  試試新東西不是比較明智嗎？

 **FAQ 6** 敲定生意的關鍵

Question

How do I answer an email message if I disagree with someone's proposal?

如果不同意某人的提案，該如何回覆？

Answer

In an email message and during meeting in English, it is important to say you disagree. One polite way is to say "I'm afraid I must disagree," or "I'm sorry but I don't agree." The important point, however, is that you give reasons for your position such as "I'm afraid I must disagree because your proposal would cost too much to implement."

不論是寫英文 email 或在英語會議中，在不同意時明確表達自身立場是很重要的。此時有幾種較有禮貌的說法，例如「I'm afraid I must disagree.」（對不起，我無法贊同。）或「I'm sorry but I don't agree.」（很抱歉，但我並不同意）。然而最重要的是，提出你採取此立場之理由，如「I'm afraid I must disagree because your proposal would cost too much to implement.」（由於執行這項提案的成本過高，很抱歉，我無法贊同。）

收到提案後，不管怎樣先回信感謝才符合禮儀。你可以事先準備好適用於所有提案的 email 範本，告知對方在公司內部討論出結果後會再回覆。

Subject: Thank you for your suggestion.

Ms. Baker,

[1]Thank you for your suggestion regarding the financial services division of our office. [2]We appreciate all input into our business and are grateful that you took the time to offer your assistance. [3]You can be assured that your suggestion will receive careful consideration. [4]If we can be of further assistance, please feel free to contact me directly.

Sincerely,

Toru Watanabe

翻譯

主旨：謝謝您的提案

Baker 小姐，您好：
感謝您對敝公司財務服務部門的相關建議。所有與敝公司業務相關之提案，我們都十分歡迎，也很感激您特地花時間提供協助。請放心，我們一定會仔細考慮您的建議。如果還有什麼我們能幫得上忙的，別客氣，請直接與我聯繫。

渡邊徹
敬啟

 重要片語 & 句型

■ **Thank you for your suggestion regarding ...** 感謝您對～的相關建議。
■ **We appreciate all input into our business.**
　　所有與敝公司業務相關之提案，我們都十分歡迎。
■ **We are grateful that you took the time to offer your assistance.**
　　很感激您特地花時間提供協助。
■ **You can be assured that your suggestion will receive careful consideration.**
　　請放心，我們一定會仔細考慮您的建議。

■ **If we can be of further assistance, please feel free to contact me directly.**
如果還有什麼我們能幫得上忙的，別客氣，請直接與我聯繫。

佳句便利貼

■ Thank you for you idea.
感謝您提出您的構想。

■ What a great idea!
這點子真棒！

■ That's not a very good idea.
這構想不太好。

■ We welcome all suggestions.
我們歡迎所有建議。

■ Rest assured, your suggestion will be considered carefully.
放心，我們一定會仔細考慮您的建議。

■ You can be certain that your suggestion will be taken seriously.
您可以放心，我們一定會慎重考慮您的建議。

■ Please feel free to call me at this number.
請別客氣，就打這支電話與我聯繫。

FAQ 7　敲定生意的關鍵

Question
What is a form letter?　　　　　何謂「form letter」？

Answer

A form letter is any letter or email message where the same letter or message is sent out to everyone. The letter may have your name at the top, or it may say "To whom it may concern," or Dear valued customer," but the content of the letter does not change from person to person. Form letters are viewed by many people as impersonal and not good business practice, but they are sometimes useful during advertising campaigns.

「form letter」是指寄給多位收件者的同內容信件或 email。這種信開頭可能寫著你的姓名，也可能寫著「To whom it may concern」（敬啟者），或「Dear valued customer」（尊貴的顧客），但是內容卻不會因為收件人不同而有所差異。許多人認為 form letter 很沒人情味，並非理想的商業手法，但就廣告活動來說，有時還挺有效的。

突顯商品的賣點
Highlighting a Product's Selling Points

推銷商品的郵件要簡潔，但應包含強調功能與價格等重要資訊。如果有圖片附檔，可能會被認為是可疑信件，故信件主旨必須小心擬定。

Subject: New! The Terra E450

Dear valued Terra customer,

[1]Please allow me to tell you about the newest product in our E-Series car lineup. [2]The Terra E450 is a top-of-the-line four wheel drive sedan loaded with everything you would expect in a Terra. [3]This car comes standard with a 7-speed manual transmission, leather seats and hand-finished interior. [4]The E450 starts at $92,500. [5]Stop by your nearest Terra dealership for a test drive.

Takuya Saito
President, Terra Motor Company of America

翻譯

主旨：新商品！Terra E450

敬愛的 Terra 產品顧客：
請容我向您介紹我們最新的 E 系列新車。Terra E450 為頂級四輪驅動轎車，搭載了所有您期待的 Terra 配備。這款車的標準配備包括 7 段式手動變速箱、真皮座椅及手工內裝等。E450 的售價 92,500 美金起。請至最近的 Terra 經銷商體驗試乘。

美國 Terra Motor 公司 總裁
齋藤拓也

 重要片語 & 句型

■ **Please allow me to tell you about** ... 請容我向您介紹～。
■ **top-of-the-line** 頂級的
■ **come standard with** ... 標準配備包括～
■ **start at** ... 從～起
■ **Stop by your nearest** ... 請至最近的～。

佳句便利貼

■ Let me highlight some of the wonderful features of our product.
請容我特別介紹本公司產品的幾項優秀特點。

■ You're going to love our newest product.
您一定會喜歡我們的最新產品。

■ Here's a product that your business is sure to need.
這正是貴公司需要的產品。

■ This product is state-of-the-art.
本商品採用最先進技術製造。

■ Standard features of our product include an energy-saving solar chip.
本產品的標準配備包括節能太陽能晶片。

■ Visit your local agent for more information.
請至您當地之代理商處索取進一步資訊。

■ Stop by a shop near you.
請前往您附近的店家洽詢。

 **FAQ 8** 敲定生意的關鍵

Question

Some sales people are very aggressive. What are some quick but polite ways to refuse a sales pitch in writing or speaking?

有些業務人員非常積極。有沒有一些快速但禮貌的書寫或口語方式來拒絕他們?

Answer

If you need to refuse a sales pitch, there are several ways you can do it. Most responses in English are very short and to the point. In email, if the sales pitch is a form letter, you can simply ignore and delete it. But, if you must answer, the polite answer is "I'm sorry but we're not interested at this time." Or you can say "We're not making any new purchases at this time." At trade shows, or when the salesperson is at your office, the best answer is "No, thanks."

要拒絕業務員推銷,有幾種方式,而大部分的英語回應都極簡短又到位。以 email 來說,若是 form letter,只要略過並刪除即可。但如果非回覆不可,較禮貌的回應為「I'm sorry but we're not interested at this time.」(很抱歉,我們目前沒興趣。)或者你可以說「We're not making any new purchases at this time.」(目前我們不打算採購任何新商品)。如果是在貿易展場或當業務員就在您辦公室裡時,最好的回答就是「No, thanks.」(不用,謝謝。)

9 針對商品提問
Inquiring about a Product

所謂與商品有關的提問也包括索取型錄或樣品等事項，而依據購買規模或金額多寡，廠商有時甚至會派銷售人員或技師來評估情況。

Subject: Brochure request

Dear Mr. Sutherland,

[1]I'm writing to you today about ABC Industries' new 7800C water pump. [2]I was wondering if you could send me a brochure about your new product or if you could have one of your sales team stop over with the information. [3]We are a medium-sized firm and are currently in the process of updating our water system. [4]Please forward the information to the address below, care of me. [5]Thank you.

Toshiyuki Nakata

Sato Components
27 Chase Boulevard
Rockton, IL 61072

翻譯

主旨：索取型錄

Sutherland 先生，您好：
今日寫這封信給您的目的是想詢問有關 ABC 工業的新 7800C 水泵產品。能否請您寄一份貴公司的新產品型錄過來，或是請一位銷售人員前來提供資訊。我們是一家中型公司，目前正在更新用水系統。請將資訊寄至下列地址，收件人寫我即可。謝謝。

中田敏行
佐藤組件
27 Chase 大道
Rockton 伊利諾州 61072

 重要片語 & 句型

■ **I'm writing to you today about ...** 今日寫信給您的目的是想～。

■ **be in the process of** ... 正在進行～

■ **Please forward the information to the address below.** 請將資訊寄至下列地址。

■ **care of** ... ～收（由～轉交）

■ **I was wondering if you could** ... 能否請您～。

 佳句便利貼

■ I'm writing to inquire about your latest product.
我來信的目的是想詢問貴公司的最新產品資訊。

■ I'd like some more information about your products.
我希望能取得更多貴公司的產品資訊。

■ Could you send me some information about your fall line?
能否請您寄一些貴公司的秋季系列商品資訊給我？

■ Could I get a brochure about your company's products?
能否提供貴公司產品型錄給我？

■ Please send it directly to me.
請直接寄給我本人。

■ Please forward the information, care of Dr. Ami Watanuki.
請將資訊寄給渡貫亞美博士，由她轉交。

FAQ 9　敲定生意的關鍵

Question

How can we write " 承蒙關照 " in English business email messages?

在英文商務 email 中，該如何表達「承蒙關照」之意？

Answer

Usually " 承蒙關照 " is translated into English as "Thank you for your continued support." But the way that it is used in Chinese is not done in English. If an English speaker wants to write "Thank you for your continued support.", then this expression would come as the closing sentence of the email message.

「承蒙關照」通常翻譯成英文的「Thank you for your continued support.」，但這句話在中文裡和英文裡用法不盡相同，如果英語人士想要表達「Thank you for your continued support.」一般會寫在郵件的結尾處。

10 推薦企業／人才
Recommending a Company / Person

當受到對方的請求、打聽而推薦、介紹公司內相關部門或相關企業，甚至是負責人員時，無論是要請對方直接與被介紹者聯繫，或表示負責者會與之聯絡，都應傳達清楚。

Subject: Our subsidiary Rising Sun Enterprises

Mr. Peter Waxman,

[1]It is a great pleasure for me to introduce to you our subsidiary, Rising Sun Enterprises. [2]Rising Sun handles research and new product development for Machida Industries. [3]I believe the project we discussed earlier would best be dealt with by this division. [4]If you desire a direct introduction to the team at Rising Sun, I would be happy to arrange that. [5]Thank you and we appreciate your business.

Sincerely,

Yuka Machida
Managing Director, Machida Industries

翻譯

主旨：我們的子公司 Rising Sun 企業

Peter Waxman 先生，您好：
很高興能為您介紹我們的子公司——Rising Sun 企業。Rising Sun 負責町田工業的研究與新商品開發工作。我相信我們之前討論的專案，由該公司負責會是最理想的。如果您希望我直接介紹 Rising Sun 的團隊給您認識，我很樂意安排。感謝您的惠顧。

町田工業 總經理
町田由香
敬上

 重要片語 & 句型

■ **It is a great pleasure for me to introduce to you** … 很高興能為您介紹～。

- **subsidiary** 子公司、相關企業
- **If you desire a direct introduction to ..., I would be happy to arrange that.**
  如果您希望我直接介紹～給您認識，我很樂意安排。
- **We appreciate your business.** 感謝您的惠顧。

佳句便利貼

- I would like to present to you our sister company Star One.
  我想為您介紹我們的姐妹公司——Star One。
- This is Yukina Kudo from our branch office.
  這位是來自我們分公司的工藤優樹菜。
- Yukari Kobayashi works for our partner firm, KY Designs.
  小林由佳里服務於我們的企業夥伴——KY 設計公司。
- This problem is best handled by our sales department.
  這問題最好能由我們的銷售團隊來處理。
- We will deal with this situation later.
  我們稍後會處理這個狀況。
- It is a pleasure doing business with you.
  很高興能與您合作。

FAQ 10 敲定生意的關鍵

Question
Should we put a seasonal or weather-related opening at the beginning of our message in English?

是否該在英文 email 開頭處，加上關於季節或天氣的問候語？

Answer
In Chinese, it is common to put a seasonal phrase, such as " 梅雨結束，即將進入夏天，不知您是否一切安好如常？", at the beginning of a business message to avoid going directly to the topic. In English, however, such expressions are not normally used. Business messages in English tend to get straight to the point.

在中文商務書信中，經常會在開頭處加上與季節相關的問候語，（例如「梅雨結束，即將進入夏天，不知您是否一切安好如常？」），以避免太直接地切入主題。但是英文並不常用這種表達方式，英文的商務 email 較傾向於單刀直入。

11 索取文件或資訊
Requesting Documents or Information

索取文件或資訊時，明確寫出目的是很重要的。請將營運狀況、時間限制、對新問題的應對等具體理由寫出來，以便索取。

Subject: Taylor and Guinness contracts

Bill,

[1]Could you please send over the contracts for the Taylor and Guinness accounts? [2]Those contracts were supposed to be delivered to this office by the 28th of the month. [3]As today is the 31st, and we are under a deadline, it is important that we move on these contacts. [4]Your prompt attention would be greatly appreciated.

Atsushi Yamada
Director of Finance, The Alenza Group

翻譯

主旨：Taylor 與 Guinness 的合約書

Bill：
能否麻煩您將與 Taylor 和 Guinness 帳戶相關的合約書寄過來？這些合約本來在本月 28 日前就應該遞交到我們辦公室的，今天是 31 號，期限快到了，我們必須趕快處理這些合約。如果您能儘速處理，我將非常感激。

Alenza 集團 財務總監
山田篤志

 重要片語 & 句型

■ **Could you please send over ...?** 能否麻煩您將～寄過來？
■ **be supposed to** V 應該要～（表示計畫或義務）
■ **be under a deadline** 期限將至
■ **Your prompt attention would be greatly appreciated.**
　如果您能儘速處理，我將非常感激。（有禮貌的催促方式）

佳句便利貼

■ Could you please expedite delivery of the layouts for the new hotel?
能否請您趕快寄出新飯店的配置圖？

■ Could you fedex the documents over immediately?
能否請您立刻將文件快遞過來？

■ This is an urgent matter.
這是急件。

■ We have a strict deadline.
我們的期限緊迫。

■ We are up against a time limit.
我們的時間非常有限。

■ It is important that we act on these contracts.
依據合約行事是很重要的。

■ We need to finalize these contracts immediately.
我們必須立刻敲定這些合約。

 FAQ 11 敲定生意的關鍵

Question
What is the difference between "could you", "would you" and "can you"?

「could you」、「would you」和「can you」的差別在哪裡？

Answer

While the usage of these three expressions is similar, the level of formality and of politeness is different. The most polite of these expressions is "could you". "Could you" will work in most business situations and is seen as very polite. "Would you" is in the middle. It is somewhat less polite, but is still used in a wide variety of business exchanges. "Can you" is less polite than the other two, but it too is used widely. The tip here is that if in doubt use "could you".

這 3 種表達的用法很類似，但正式度和禮貌度則有所不同。其中最有禮貌的是「could you」，這在大多數商務場合都適用，且被視為極有禮貌的講法。「would you」則中等，禮貌度低一點，但仍廣泛被運用於各式商務往來場合中。「can you」則沒有另外兩者那麼禮貌，但是使用範圍也很廣。如果不確定該用何者時，用「could you」是最保險的。

想知道業務進展狀況時，只要用以下範例中的「Would you mind giving us an update on... ?」來切入即可。英文 email 最重要的就是先把目的寫清楚，並不需要什麼開場白。

Subject: Update request

Mr. Rogers,

[1]Would you mind giving us an update on the upgrades to our computer system? [2]Specifically, when will the work be finished, and more importantly, will there be any more disruptions to our network? [3]Any information you could provide would be greatly appreciated. [4]Please respond at your earliest convenience. [5]Thanks.

Atsushi Yamada
President, Yamatake Hands

翻譯

主旨：請提供最新狀況

Rogers 先生，您好：
能否請您提供我們電腦系統升級作業的最新狀況？明確地說，你們何時能完成？更重要的是，我們的網路還會發生任何問題嗎？只要你們提供任何資訊，我都會非常感激。請儘快回覆，謝謝。

山武手創館 總裁
山田篤志

 重要片語 & 句型

- **Would you mind giving us an update on ...?** 能否請您提供～的最新狀況？
- **specifically** 明確地說
- **and more importantly** 更重要的是
- **Any information you could provide would be greatly appreciated.**
  只要你們提供任何資訊，我都會非常感激。
- **Please respond at your earliest convenience.** 請儘快回覆。

佳句便利貼

■ Would you take a moment to answer a few questions?
能否請您花點時間回答幾個問題？

■ If you have any information, could you please let me know?
如果您有任何訊息，能否讓我知道？

■ The problem is namely about money.
問題就在資金上。

■ Please get back to me when you have a chance.
有空時請回覆我。

■ Please let me know ASAP.
請儘快通知我。

 FAQ 12 敲定生意的關鍵

Question
How do I introduce myself to someone in an email message or when I first make contact by email?

我該怎麼用 email 把自己介紹給對方，或是第一次用 email 與對方聯繫時該怎麼寫？

Answer
To introduce yourself or to make first contact with someone at a different company, you begin with "Dear Mr./Ms./Mrs." and their family name. Do not use first names until you are better acquainted with the person. Then, the first line of the letter should begin "My name is (first name, last name) and I am (position) at (company name)". For example, "My name is Tatsuya Mori and I am the managing director of Sakura One Bank". Once you have done this, then you can move on to the reason you are writing to them.

如果要介紹自己，或是首度與其他公司裡的某人聯繫時，應以「Dear Mr./Ms./Mrs.」加上對方姓氏來開頭。除非已跟對方很熟，否則千萬別直呼其名。然後第一行應該要寫「My name is（名字，姓氏）and I am（職稱）at（公司名稱）.」。例如「My name is Tatsuya Mori and I am the managing director of Sakura One Bank.」（我叫森達也，是 Sakura One 銀行的總經理）。寫完上述句子後，就可陳述你來信的原因了。

13 確認庫存
Checking Stock

收到大量訂單時，要先確認庫存與生產時程才能回覆客戶。負責接受訂單的人員應培養正確的生產管理知識，並時時掌握庫存狀況以便應對。

Subject: Inventory check request

Bill,

[1]Please check our inventory in the warehouse to see how many D-345 engines we have on the shelves. [2]We have a client who would like to make a major purchase of 16 engines and we need to know how many are in stock and how many will need to be made. [3]As this is a priority, please get back to me ASAP. [4]Thanks.

Mie

翻譯

主旨：煩請確認庫存

Bill：

請確認我們的倉庫庫存，看看架上還有多少 D-345 引擎。有個客戶想要大量採買，共需要 16 台引擎，我們必須了解有多少庫存，也得知道該再生產幾台。這件事必須優先處理，所以請儘快回覆我。謝謝。

美惠

 重要片語 & 句型

- **Please check our inventory in the warehouse to see how many ... we have on the shelves.** 請確認我們的倉庫庫存，看看架上還有多少～。
- **We have a client who would like to V.** 有個客戶想要～。
- **We need to know how many are in stock.** 我們必須了解有多少庫存。
- **As this is a priority, please get back to me ASAP.**
  這件事必須優先處理，所以請儘快回覆我。

佳句便利貼

- Please have a look at what's in storage.
  請確認一下庫存現在有什麼。

- Please look in the closet.
  請查看一下儲藏室。

- A customer has requested seven bottles of our new tonic.
  有位客戶訂了 7 瓶我們的新款護髮劑。

- How many are available?
  庫存有多少？

- As this is an urgent matter, please respond quickly.
  這是急件，請趕快回覆。

- Make this your priority.
  請優先處理這件事。

 FAQ 13 敲定生意的關鍵

Question

When can I use someone's first name in a business email?

什麼時候才可於商務 email 中直呼某人的名字（first name）？

Answer

Using the other person's first name sometimes depends on the situation. If they are in a subordinate position, it is possible to use their first name. If it is someone at another company, it depends on how well you know them. A good rule to follow is that if the other person uses your first name, then it is probably safe to use their first name in email correspondence.

是否直接稱呼他人名字有時需視情況而定。若是自己的屬下，可直呼其名；若是其他公司的人，就要看你們彼此有多熟了。而最好的判斷方法，就是如果對方直接稱呼你名字，便代表你也可以在 email 往來中直接稱呼對方的名字。

14 聯繫某個部門或團隊
Contacting a Department or Team

現在這個時代，公司內部聯繫基本上也都使用 email。「忘了看信」這種藉口根本不被接受。另外，發信者不該散彈槍式地傳送多封 email，而應把相關資訊加以統整後，再一次寄出。

Subject: Team merger announcement

Product development team members,

[1]I would like to bring to your attention the memo I received from our corporate headquarters. [2]In it, the president of the company, Mr. Yukio Iizuka, announced that the product development team will be merged with the research team to form a new joint group. [3]This will be effective from September 1st. [4]Any questions may be directed to me.

Shota Kudo

翻譯

主旨：部門合併通知

產品開發部的夥伴們：
在此要告知各位一則我從總公司收到的備忘錄。其中提到，本公司總裁飯塚幸雄先生宣布，產品開發團隊將與研究團隊合併為新的聯合部門。此項合併將於 9 月 1 日起生效。如果有任何疑問，請直接與我聯繫。

工藤翔太

 重要片語 & 句型

■ **I would like to bring ... to your attention.** 在此要告知各位～。
■ **This will be effective from ...** 這將於～起生效。
■ **Any questions may be directed to me.** 如果有任何疑問，請直接與我聯繫。

佳句便利貼

■ This policy is effective immediately.
此項政策即刻生效。

■ Our new program will go into effect in August.
我們的新計畫將於 8 月開始實施。

■ Direct all questions to our lawyers.
所有問題都請直接向本公司律師（法律顧問）提出。

■ Questions about the program may be forwarded to my secretary.
關於此計畫的所有問題都請傳給我的秘書。

 FAQ 14 敲定生意的關鍵

Question
How do I express " 各位 " in English?　　如何用英文表達「各位」之意？

Answer

You can use "To all (name of section or department) staff [members]," as in "To all sales staff," or "To all R&D team members," This clearly means the message is meant for everyone in that part of the company.

你可以用「To all（部門或單位名稱）staff (members)」這樣的方式來寫，例如「To all sales staff」或「To all R&D team members」。這樣就能清楚表示此信是給公司該部門的所有人看的。

企業間在做交易時，只要訂購商品一送達，通常就必須立刻聯絡對方。比電話更能精確地告知對方並且能留下文字記錄之特性，正是 email 的優勢所在。

Subject: Parts received

Mr. Bill Prince,

[1]This message is to acknowledge receipt of the parts shipment order #84123, dated October 10, 20XX. [2]We have inspected all of the parts and would like to thank you for your prompt shipment of our order. [3]We look forward to doing business with you again in the future.

Sincerely,

Masato Kogame
Procurement Officer, Matsu Fabrication Inc.

翻譯

主旨：零件已收到

Bill Prince 先生，您好：
本信的目的是要通知您，我們已收到日期為 20XX 年 10 月 10 日，訂單編號 #84123 之零件。我們已檢查過所有零件，並要感謝您如此迅速地將我們訂購的貨品送到。期待未來能再次與您合作。

松製造公司 採購員
小龜正人
敬啟

重要片語 & 句型

■ **This message is to acknowledge receipt of ...** 本信的目的是要通知您，我們已收到～。
■ **We would like to thank you for your prompt shipment of our order.**
　感謝您如此迅速地將我們訂購的貨品送到。
■ **We look forward to doing business with you again.** 期待未來能再次與您合作。

佳句便利貼

■ I'm writing to let you know that our order hasn't arrived.
寫這封信的目的是要通知您，我們訂購的商品尚未送達。

■ Thank you for quickly sending the required documents.
謝謝您迅速寄來所需文件。

■ We are certain to do business with you again.
我們一定會再次合作的。

■ We welcome the opportunity to work with you in the future.
我們很期待將來有機會與您合作。

 FAQ 15 敲定生意的關鍵

Question
What should I do if I can't tell from the person's name if they are a man or a woman?

無法從對方的名字判斷出是男是女時，該怎麼辦？

Answer
In order not to offend someone such as calling a woman Mr. or a man Ms., it is best to try to find out if the person you are writing to is male or female. Doing an Internet search will sometimes give you the answer or you could call the head office and simply ask if the person you wish to contact is male or female. In this way, you eliminate a potentially embarrassing situation.

為了避免冒犯對方，比如將女性稱為 Mr.，或將男性稱為 Ms.，最好先試著查出你寫信的對象到底是男是女。有時搜尋網路就能找到答案，或者你也可以打電話給總公司，直接詢問欲聯繫對象的性別，這樣就能避免可能發生的尷尬場面。

3

合約・訂單
Contract / Order

如果是未刊載於型錄上的商品、服務，或因訂購特殊組合而不確定價格如何時，通常會先用 email 請對方報價，再依據預算來決定是否購買。

Subject: Estimate request

Hi Jane,

[1]This is Saeko Shimada over at ABC Supply. [2]I was wondering if we could get an estimate for a D-130 paint gun and air compressor. [3]Please include the cost of delivery as we would like it sent over to our shop. [4]If you could get in touch with me when you have the quote, I would appreciate it.

Thanks,

Saeko Shimada
Supply Manager, ABC Supply

翻譯

主旨：請求報價

嗨，Jane：

我是 ABC 器材的島田紗枝子。不知您是否可提供 D-130 油漆噴槍與空氣壓縮機的報價給我們？請將運費也納入，因為我們希望能直接送到店裡來。如果您報價單準備好了就請與我聯繫，我將十分感激。

謝謝。
ABC 器材 採購經理
島田紗枝子

 重要片語 & 句型

■ **This is A over at B.** 我是 B 的 A。(此句型常用於商務 email 開頭的自我介紹)
■ **I was wondering if we could get an estimate for ...** 不知您是否可提供～的報價給我們？
■ **Please include the cost of delivery.** 請將運費也納入。
■ **would like A sent over to B** 希望能將 A 直接送到 B 處
■ **If you could get in touch with me when you have the quote, I would appreciate it.**

46　　如果您報價單準備好了就請與我聯繫，我將十分感激。

佳句便利貼

- Could we get an estimate on the cost of repairs?
  可否提供我們修理費的報價？

- Could you provide us with a rough idea of the cost?
  能否請您概估一下費用？

- Please figure in the shipping cost.
  請將運費也納入。

- Don't forget the mailing costs.
  別忘了郵資。

- Mail me when you've got the estimate.
  報價單準備好後，請寄 email 給我。

- Please give me a call.
  請打電話給我。

FAQ 16　敲定生意的關鍵

Question
What if someone I don't know uses my first name in an email message to me and it makes me uncomfortable?

如果有某個我不認識的人在 email 裡直接稱呼我名字而讓我不太自在，這時該怎麼辦？

Answer
Please remember that in dealings with English-speaking cultures, the use of the first name is common. It is not used disrespectfully. Rather, it is a way of treating the other person as an equal rather than simply as a customer or acquaintance. If you deal with English speakers long enough, it will become natural to use the first name in correspondence.

請記得，在英語文化中直呼對方名字是很常見的事，並無不尊重之意。相反地，這表示對方將你視為平等對象，而非單純的客戶或點頭之交。如果你和英語系國家的人相處夠久，就會很自然地以名字互稱。

依據行業別、時期、購買數量及以往的購買紀錄，賣方有時會願意降價銷售。此時買方不應逼迫對方打折，而必須好好說明原因，例如是受預算限制之故，或是大宗採購對對方有何好處等。

Subject: A-330 golf cart question

Dear Allied Products,

[1]Thank you for sending me the catalog of your products. [2]I'm particularly interested in the A-330 golf cart, but the price is slightly over budget. [3]I was wondering if there are any discounts available, or will it be going on sale soon? [4]Please let me know if there are, and perhaps we can work out a deal. [5]Thanks.

Tadashi Yoshida
Pacific Country Club

翻譯

主旨：有關 A-330 高爾夫球車的問題

Allied Products 公司：
感謝您寄來貴公司商品型錄。我對 A-330 高爾夫球車特別有興趣，但是價格稍微超出了敝公司預算。不知是否有任何折扣方案？或是近期將有優惠活動？如果有相關訊息，煩請通知，或許我們可以做成一筆生意。謝謝。

太平洋鄉村俱樂部
吉田正志

 重要片語 & 句型

■ **Thank you for sending me ...** 感謝您寄來～。
■ **I'm particularly interested in ...** 我對～特別有興趣。
■ **The price is slightly over budget.** 價格稍微超出了預算。
■ **I was wondering if there are any discounts available.** 不知是否有任何折扣方案？
■ **Perhaps we can work out a deal.** 或許我們可以做成一筆生意。

■ That is out of our price range.
該價格超出了我們的預算。

■ It is more than we can afford.
超過我們所能負擔。

■ Can we get a discount?
可以打個折嗎？

■ Can you give us a deal on this?
這部分可以打個折嗎？

■ Perhaps we can work something out.
或許我們能找出解決辦法。

■ I'm very interested in your new sales idea.
我對貴公司的新銷售案很有興趣。

■ We are becoming increasingly interested in a tie-up between our two companies.
我們對於貴公司與敝公司的合作愈來愈感興趣。

 FAQ 17 敲定生意的關鍵

Question

What should I say if I accidentally send an email message to the wrong person at the wrong company?

如果不小心寄錯 email 給不對的公司、不對的人，該怎麼辦？

Answer

Sometimes we mistakenly send an email message to someone when the message was meant for someone else. If it is someone you have frequent contact with, then simply send a quick message saying "Please disregard the previous message. It was sent to you by mistake." If it is someone from another company, or someone you have less contact with, then you should say the same thing, but add a short apology at the end such as "Sorry about that."

我們偶爾會寄錯 email 給不對的收件者。如果這個收件人是你常聯繫的對象，那麼只需趕快寄一封簡短的 email，寫上「Please disregard the previous message. It was sent to you by mistake.」（請忽略上一封信，我寄錯人了。）而如果這位收件人是在其他公司、你不常聯絡的人，那麼你也應該發出同樣內容的 email，但是還需在末尾做個簡短致歉，如「Sorry about that.」（這件事很抱歉。）

18

提出折衷價格
Offering a Compromise Price

議價到最後，如果還想再殺價，就該具體提出你的期望價。另外，如下例所示，如果能清楚寫上可立即付款，對於議價來說是很有利的。

Subject: Forklift offer

Arnold,

[1]Thanks for getting back to me about the price of the used forklift. [2]Your price of $2,000 is a bit over our spending limit of $1,800. [3]Would you be prepared to accept $1,800 for it? [4]If that price is acceptable, we will forward the money to your account immediately. [5]Please let me know.

Thanks,

Hiroyuki Sasamura
Koala Moving Company

翻譯

主旨：堆高機的出價

Arnold：
謝謝您對中古堆高機價格的回覆。您提出的 2,000 美金稍微超出了敝公司預算限制的 1,800 美金。您可否接受 1,800 美金的價格？如果可以的話，我們會立即匯款至貴公司帳戶。請再告知我您的決定。

謝謝。
無尾熊搬家公司
笹村博之

 重要片語 & 句型

■ **Thanks for getting back to me about ...** 謝謝您對～的回覆。
■ **Your price of ... is a bit over our spending limit of ...**
　 您提出的～價格稍微超出了敝公司預算限制的～。
■ **Would you be prepared to V?** 您可否接受～？
■ **forward the money to** *one's* **account immediately** 立即匯款至某人的帳戶

佳句便利貼

■ Thanks for mailing me back.
感謝您的回信。

■ Thanks for the quick response.
感謝您迅速回覆。

■ Our budget for the new system is $10,000.
我們新系統的預算為 10,000 美金。

■ How about $950 for the database?
這資料庫賣 950 美金如何？

■ Would you accept $1,400?
1,400 美金可以接受嗎？

■ If we are in agreement on the price, I can send over the payment tonight.
如果同意這個價格，那我今晚就可付款。

■ If our offer is acceptable to you, please let me know.
如果您能接受我們的出價，就請通知我一聲。

 FAQ 18 敲定生意的關鍵

Question

What if one of our customers sends us a private email by mistake? Should I mention it to them?

如果有客戶誤將私人 email 寄來，我該跟他們說嗎？

Answer

In situations where the client or customer has sent a potentially embarrassing email message to you by mistake, the best thing to do is simply delete it and not mention it to the client unless the client mentions it first. If they do, then a simple response of "No problem." will help overcome the situation.

如果有客戶誤將可能造成尷尬的私人 email 寄給你，此時最好的辦法就是刪除該信，且不要主動告知客戶，除非該客戶先提起。如果客戶提到此事，也只需簡單回應一句「No problem.」（別擔心。），就能順利解決狀況。

19 交渉合約內容
Negotiating a Contract

合約內容是該直接面對面交涉，或是透過電話討論，甚至只用 email 往返，需依個案情況而定。以下範例則明確指出了合約問題所在。

Subject: About the contract

Mr. Wayne Whitehouse,

[1]Thank you for sending me over the proposed contract for the plumbing work at our new business location. [2]My lawyers have looked over it and there are a couple of minor details that I would like to discuss with you, namely Articles 13 and 15 of the contract. [3]If you could give me a call, I'm sure we can come to an agreement and finalize the contracts.

Mitsuaki Sano
CEO, Takeda Drills

翻譯

主旨：關於合約

Wayne Whitehouse 先生：
感謝您寄來有關敝公司新辦公室水管工程的合約書提案。我的律師已看過該合約書，其中有些小細節（即合約中的第 13 與 15 條），我想與您討論一下。如果您能來電，我想我們一定能達成協議並敲定合約。

竹田鑽具 CEO
佐野光昭

 重要片語 & 句型

■ **Thank you for sending over the proposed contract for ...**
感謝您寄來有關～的合約書提案。

■ **look over ...** 查看～

■ **There are a couple of minor details that I would like to discuss with you.**
有些小細節我想與您討論一下。

■ **I'm sure we can come to an agreement and finalize the contracts.**
我想我們一定能達成協議並敲定合約。

佳句便利貼

■ I'm reading the contract sample you sent over last week.
我正在讀您上週寄來的合約樣本。

■ This is our standard contract.
這是我們的標準合約。

■ Management has examined the documents.
管理階層已檢視過這些文件。

■ I will read through the contracts to see if everything is in order.
我會詳讀合約，看看一切是否妥當。

■ I'm sure we can reach an agreement.
我想我們一定能達成協議。

■ I'm sure we can resolve these minor points.
我想我們一定能解決這些小問題。

■ I'm sure we can find common ground.
我想我們一定能找出彼此都能接受的共通點。

FAQ 19　敲定生意的關鍵

Question

How can I make sure my email is read and doesn't end up in someone's junk mailbox?

我要怎麼確定發出去的 email 確實有被讀取，而不至淪落到某人的垃圾郵件匣中？

Answer

If you are contacting someone for the first time, your email may be sent directly to that person's junk mailbox, where they may or may not see it. One of the best ways to avoid this is to use a clear subject line, stating what the email is about. For example, "SY Ltd. Contracts" or "Follow-up to our meeting last week: Maiko Kaneko." While no guarantee your email will be read, it may make your message stand out and not to be mistaken for junk mail. A second way is to email them through a link on their website.

如果是第一次與該對象聯繫，那你的 email 很可能會直接被轉到對方的垃圾郵件匣中，故對方可能會看到也可能不會看到。要避免這種情況發生最好的辦法之一，就是寫上清楚的信件主旨，講明此信之目的。例如「SY Ltd. Contracts」(SY 公司合約) 或者是「Follow-up to our meeting last week: Maiko Kaneko.」(上週會議的後續聯繫：金子麻衣)。雖然無法保證對方一定會讀到你的信，但明確的主旨可讓你的信件較顯眼，且不至於被誤認為垃圾郵件。另一種方式則是透過他們網站上的連結來寄送 email。

同意合約內容
Agreeing to a Contract

雙方都同意合約內容後，通常就會將合約做成正式文件。準備好一式兩份的合約文件後，經雙方聯合署名，再各自保管一份。合約書與一般 email 不同，必須非常注意不能發生任何英文上的錯誤。

Subject: Contract completion

Ms. Ronda Harrington,

[1]We are pleased to announce that we have looked over the contracts you sent over last week and everything is in order. [2]We have signed the contracts and will send back one set of contracts to you as requested, and will keep the second copy for our records. [3]We welcome the opportunity for this tie-up between Harrington Fashions and Chrysalis Shirt Company. [4]We hope the relationship will be mutually beneficial and look forward to a long working relationship together.

Sincerely,

Takeshi Wada
Owner, Chrysalis Shirt Company

翻譯

主旨：簽約完成

Ronda Harrington 小姐：
很高興通知您，我們已詳閱過您上週寄來的合約，且確定一切都沒問題。我們已簽好合約，並且會依照您的要求，將其中一份寄回給您，另一份則留存本公司。很高興 Harrington 流行服飾公司與蝶蛹服飾公司能有機會密切結合，希望這份關係能讓彼此互利，同時也期待我們能有長期性的合作。

蝶蛹服飾公司 業主
和田高志
敬啓

- **We are pleased to announce that ...** 很高興通知您～。
- **Everything is in order.** 一切都沒問題。
- **as requested** 依照要求
- **We will keep the second copy for our records.** 另一份則留存本公司。
- **We look forward to a long working relationship together.**
  我們很期待能有長期性的合作關係。

佳句便利貼

- Everything appears to be correct.
  一切看來都沒問題。

- Everyone has signed off on your proposal.
  每個人都簽署了您的提案。

- We look forward to meeting with you in person next week.
  我們很期待下週與您見面。

- We think both companies will benefit from this line of products.
  我們認為兩家公司都能因為此商品系列而獲益。

- We believe this new office design will benefit us mutually.
  我們相信這個新的辦公室設計對雙方都有利。

 FAQ 20 | 敲定生意的關鍵

Question
When can I use "Sir" or "Madam?"　　　何時該用「Sir」或「Madam」？

Answer

"Sir" and "Madam" are becoming increasingly rare in email messages. But it is possible to use them in certain cases. For example, if you are making first contact by email through a website link and don't know who your message is going to in the company, then you may want to start your email message "Dear Sir or Madam,." Another time is when you are sending an email message to a group (such as a group of male lawyers). In this case, use the plural sirs, as in "Dear Sirs,"

在電子郵件中,「Sir」或「Madam」的使用率愈來愈低,但是在某些情況下仍可使用。例如,如果你是透過網站連結首度以 email 與對方聯繫,而不知道信會寄到該公司什麼人的手上時,就可在 email 中以「Dear Sir or Madam,」起頭。另一種情況則是寄 email 給一群人時(例如一群男性律師),這種時候便可用複數的 sirs,如「Dear Sirs,」。

21 修改合約內容
Modifying a Contract

基本上，在合約期間就該遵守合約內容。如果要更改合約內容，則須在更新合約時處理。以下便是企業（非個人）要求更改手機合約內容（更改手機數量及計費方案等）的例子。

Subject: About our cell phone contract

Dear Mr. Riley,

[1]Thank you for notifying us that our current cell phone contract is about to expire. [2]We would like to make a few changes to our current contract to better reflect our company's needs. [3]Could you please drop by our office when you are available to discuss the changes that are necessary for our new contract? [4]Thank you.

Yasunori Ohno
Ohno Cement

翻譯

主旨：關於我們的手機合約

Riley 先生，您好：
感謝您告知我們目前的手機合約即將到期。我們打算針對現有合約做些修改，以符合敝公司的需求。能否請您擇空前來敝公司，以便討論新合約所需更動之事項？謝謝。

大野水泥公司
大野康則

 重要片語 & 句型

■ **Thank you for notifying us that ...** 感謝您告知我們～。
■ **be about to expire** 即將到期
■ **We would like to make a few changes to our current contract.**
　我們打算針對現有合約做些修改。
■ **reflect** *one's* **needs** 符合～的需求

56

■ **Could you please drop by our office when you are available to** …
能否請您擇空前來敝公司，以便～。

佳句便利貼

■ Table 3 of your prospectus better shows the new building design.
說明書中的表 3 更清楚地展示出新大樓的設計。

■ Let's discuss what needs to be done to improve our bottom line.
讓我們討論一下該怎麼做才能改善收益。

■ These parts of the contract need to be fixed.
合約中的這些部分必須修改。

■ Please stop in when you have a moment.
有空時請過來一下。

 **FAQ 21** 敲定生意的關鍵

Question
If I send a short email message, won't the recipient think I am unprofessional?

如果我寄的 email 很簡短，對方會不會覺得我不專業？

Answer
Not at all. In business, people are busy and they don't have a lot of time to deal with all of the mail coming in. People appreciate it if you are brief and to the point, and it will make the reader's job easier.

一點都不會。在商務環境中大家都很忙，沒太多時間處理每一封信。一般人對於寫得簡短扼要的 email 都心懷感激，因為這能讓讀信者的工作更輕鬆。

提出折衷方案
Offering a Compromise on a Contract

商務合約以當事人彼此達成協議為前提，並無第三者居中協調，故當雙方對合約內容意見相左時，必須有一方表示妥協，否則合作必定破局。

Subject: Contract resolution

Steve,

[1]We seem to have reached a deadlock on the contracts. [2]So, in order to move things forward and get this contract completed, we are willing to compromise on the price of $2.32 per square meter of flooring as you proposed. [3]Please send the contracts over and we will get them signed and back to you as soon as possible.

Thank you.

Shin Nakamura
Flooring Masters

翻譯

主旨：合約解決方案

Steve：

看來我們在合約方面已陷入僵局。所以，為了讓事情能繼續進行並完成簽約，我們願意妥協，接受您所提出每平方公尺 2.32 美金的地板價格。請將合約寄過來，我們會儘快簽好並寄回給您。

謝謝。
地板大師公司
中村 真

重要片語 & 句型

■ **We seem to have reached a deadlock on ...** 看來我們在～方面已陷入僵局。

■ **in order to move things forward** 為了讓事情能繼續進行

■ **We are willing to compromise on the price of ...** 我們願意妥協，接受～的價格。

■ **Please send the contracts over and we will get them signed and back to you as soon as possible.** 請將合約寄過來，我們會儘快簽好並寄回給您。

佳句便利貼

■ We seem to have hit a roadblock in our construction plans.
看來我們的建設計畫遇到了障礙。

■ This bill is stuck in committee.
此議案卡在委員會中。

■ In order to get things moving again, let's meet to discuss the deal tonight.
為了使事情再度有所進展，讓我們今晚會面討論一下這項交易。

■ In order to take the next step, we first need to complete this paperwork.
為了進入下一階段，我們必須先完成這份文件。

■ We would like to offer up a compromise in order to get your business.
為了達成交易，我們願意妥協。

■ We have done nothing but compromise, but you seem unwilling to do the same. 我們不斷妥協，但是您似乎不為所動。

■ We will have the signed contracts sent over this evening.
我們今晚會將簽好的合約寄出。

 **FAQ 22** 敲定生意的關鍵

Question
How do I bargain with a seller, for example, in an online auction?

我該怎麼跟賣家討價還價，比如說，在拍賣網站上？

Answer
Online auctions can bring great savings for your business if you can get a bargain. One of the best ways is to send the seller an email message outlining your offer. You could write, "Instead of $225, how about this deal? I will give you $200+postage and will pay by Paypal immediately upon receiving your agreement. If this is acceptable, please let me know." Very often this will be successful.

如果能議價成功，網路拍賣確實可幫你的公司省下不少錢。而討價還價的最好辦法就是寄信給賣家，告訴對方你想出價多少。你可以這樣寫：「Instead of $225, how about this deal? I will give you $200+postage, and will pay by Paypal immediately upon receiving your agreement. If this is acceptable, please let me know.」（不要 255 美金，這樣如何？我給你 200 美金加上郵資，並在您同意後立刻以 Paypal 支付。若您覺得可行，就請通知我一聲。）通常這樣都會成功。

簽約以後,也可能因為某些理由而解約。有些是在合約期間中途解約,但大部分的情況是合約過期失效後不再續約。無論是什麼原因,都應向對方清楚說明理由。

Subject: Contract cancelation

Dr. John Mathews,

[1]This is to inform you that we have decided not to renew our bottled water contract with you upon expiration on June 30th. [2]Unfortunately, our company is going through a bad economic situation and these cutbacks are necessary. [3]Once the economy improves, we may again contact you. [4]Thank you.

Mio Kondo
Chief Financial Officer, Kondo Dental

翻譯

主旨:解除合約

John Mathews 博士:
本信的目的是要通知您,我們決定在 6 月 30 日合約到期後,就不再更新與貴公司簽訂的瓶裝水合約。很不幸,敝公司目前正處於營收不佳期間,必須削減這些支出。一旦我們的經濟情況好轉,或許會再重新與您聯絡。謝謝。

近藤牙醫 財務總監
近藤美緒

 重要片語 & 句型

■ **This is to inform you that** ... 本信的目的是要通知您~。
■ **renew** *one's* **contract with** ... 更新與~的合約
■ **upon expiration** 一到期
■ **go through** ... 經歷~
■ **once the economy improves** 一旦經濟情況好轉

佳句便利貼

■ We are cancelling the contract.
我們要取消合約。

■ We would like to nullify this contract.
我們想廢除此合約。

■ We must make significant cuts to our budget.
我們必須大幅削減預算。

■ The head office is looking to cut expenses wherever possible.
總公司希望能儘量削減支出。

■ Once business picks up, we will begin hiring new employees.
一旦生意好轉，我們就會開始雇用新人。

 FAQ 23 敲定生意的關鍵

Question

Isn't it bad to give a reason why we are cancelling a contract?

跟對方解釋為何取消合約，不會不恰當嗎？

Answer

Not at all. Just as with giving a reason when you cancel a meeting or are late, you should also give a reason why you are cancelling a contract. Your answer may help the other company improve their services. In many English-speaking countries, there is an expectation of a reason. Not giving a reason seems cold.

一點也不會。就跟取消會議或遲到時要給理由一樣，取消合約時也應該提出原因。你的理由或許還能幫助對方提升其服務品質。在許多英語系國家，通常會希望了解原因，不說明白反而會讓人覺得很冷淡。

合約書多半會在雙方同意草案內容之後，才正式擬定為文件。有時會先請對方確認草稿，再準備好正式版本；有時則不經這道手續，直接做好合約書並寄給對方簽名。

Subject: Contracts sent

Deb,

[1]I just sent over the completed contracts for our new distribution agreement. [2]They should get to you by Friday. [3]Please look them over upon receipt and if there are any problems, let me know. [4]Thanks and we look forward to working with you.

Asami Sudo
Bando Bottling, Ltd.

翻譯

主旨：合約已送出

Deb：
我剛剛把已完成的新版分銷協議合約寄過去了，週五前應該就會寄到。收到後請詳閱合約，如果有任何問題，請通知我一聲。感謝您，同時很期待與貴公司合作。

坂東裝瓶公司
須藤麻美

 重要片語 & 句型

- **I just sent over the completed contracts for ...** 我剛剛把已完成的～合約寄過去了。
- **They should get to you by ...** ～前應該就會寄到。
- **look ... over** 詳閱～
- **upon receipt** 一收到
- **If there are any problems, let me know.** 如果有任何問題，請通知我一聲。
- **We look forward to working with you.** 我們很期待能與貴公司合作。

佳句便利貼

■ I just mailed you the contract sample you requested.
我剛把你要的合約樣本寄去給你了。

■ The contracts are in the mail.
合約已郵遞出去。

■ The contracts should reach you shortly.
合約書應該很快就會送到。

■ Please read over the contents of the contract.
請詳讀合約內容。

 FAQ 24 敲定生意的關鍵

Question

How do I write " 辛苦你了 " in a message when someone has completed a project?

如何以 email 對完成專案的人表達「辛苦你了」之意？

Answer

" 辛苦你了 " doesn't translate directly. Instead we look at how " 辛苦你了 " is used to see if there is a similar usage in English. At the end of a project, one might say in English "Good work!" or "Good job!" or "Thank you for your hard work." Each of these represents the meaning and feel of " 辛苦你了 ."

一般不會將「辛苦你了」直譯成英文，而是依據「辛苦你了」這句話的使用情境，來判斷在英文中是否有類似說法。例如在專案即將結束時，有些人可能會用英文說「Good work!」、「Good job!」（做得好！）或「Thank you for your hard work.」（謝謝你的努力。）等，而這些說法都能傳達出「辛苦你了」的意義與感覺。

25 訂購商品（下單）
Ordering a Product

在販賣商品的企業網站上確認過圖片、價格、功能後下單購買，並以信用卡結帳的方式已相當普遍。不過在確定購買前應先用 email 或電話確認不清楚或有疑慮的部分。

Subject: New order

Greetings,

[1]I would like to order the following: Four all-season radial tires for each of the eight 20XX Skylarks in our corporate fleet. [2]According to your website, these are currently on sale for $899 per set of four. [3]Could you confirm that you have enough in your inventory to supply our company? [4]Thank you and your prompt response would be greatly appreciated.

Yuko Yokoi
JP Transport

翻譯

主旨：新訂單

您好：

我想訂購以下商品：本公司車隊共 8 輛 20XX 年 Skylark，每台需要 4 個四季型輻射層輪胎。依據貴公司網站資料，這些輪胎目前 4 個一組特價 899 美元。能否請您確認現有庫存是否足夠提供給敝公司？謝謝。如果能儘快收到您的回覆，我將萬分感激。

JP 運輸公司
橫井優子

 重要片語 & 句型

■ **I would like to order the following:** 我想訂購以下商品：（Note: 用冒號）
■ **be currently on sale for ...** 目前特價～（金額）
■ **Could you confirm that you have enough in your inventory?**
　能否請您確認現有庫存是否足夠？
■ **Your prompt response would be greatly appreciated.**
　如果能儘快收到您的回覆，我將萬分感激。

■ I wish to order this desk.
我想訂購這張桌子。

■ Can I place an order?
我可以下單嗎？

■ Could you make sure you have this supplement in stock?
能否請您確認一下此商品庫存是否足夠？

■ Are you certain that this work station is still available?
您確定這組電腦工作站還有貨嗎？

■ These tools are being discounted at the moment.
這些工具目前有折扣。

■ There is a sale going on now.
現在有特價活動。

FAQ 25 敲定生意的關鍵

Question

How can I ask for clarification if I receive an email message from a client that I don't understand?

如果收到一封客戶的 email，但是看不懂內容，我該如何請對方說明？

Answer

Sometimes what is written in a message is not clear to the recipient. If this is the case, you can use phrases like "I'm afraid I don't understand your question." or "Could you please clarify what you mean by …?" This will usually solve the problem.

有時 email 的內容對收件者來說可能語焉不詳。在這種情況下，你可以用「I'm afraid I don't understand your question.」（我恐怕不太了解您的問題。），或「Could you please clarify what you mean by...?」（能否請您說明一下～的意思？）這類句子回覆。這樣通常就能解決問題。

下單後，偶爾還會需要增加訂購量，或是加購其他商品。而一般來說，統一運送才能節省運費，因此如果不放心，可直接打電話與對方聯繫、確認。

Subject: Order change request

Alonzo,

[1]About the order I sent you this morning, is there still time to change it? [2]I would like to add two more boxes of photocopiers paper (from four boxes to six). [3]If the order is already on the truck and en route, then please disregard and I will hold off until next time. [4]Anyway, let me know. [5]Thanks.

Kohei
Wild Entertainment

翻譯

主旨：麻煩修改訂單

Alonzo：
我早上寄給您的訂單還來得及改嗎？我想再追加 2 盒影印紙（從 4 盒改為 6 盒）。如果該筆訂單已裝車出貨，就請忽略此信，我會延至下次再訂購。無論如何，請通知我一聲。謝謝。

荒野娛樂公司
耕平

 重要片語 & 句型

■ **Is there still time to V?** 還來得及～嗎？
■ **I would like to add ...** 我想再追加～。
■ **be already on the truck and en route** 已裝上卡車運送中
■ **disregard** 忽略
■ **I will hold off until next time.** 我會延至下次。

佳句便利貼

■ I'm checking on this morning's order.
我正在確認今早的訂單。

■ Do you remember that order I sent you?
還記得我寄給您的訂單嗎?

■ Do we have time to adjust that order?
還來得及調整那份訂單嗎?

■ We're running out of time.
我們快沒時間了。

■ Please ignore my last message.
請忽略我的前一封信。

■ We will delay making a decision until after next week's meeting.
我們將延至下週會議後再作決定。

 FAQ 26 敲定生意的關鍵

Question

Should I send a confirmation message when my colleague sends me work or has a request?

當同事寄來工作相關或請求協助的 email 時,我是否應該回覆確認信?

Answer

Definitely. Just a quick note that you have received the work will put the sender at ease. You might say, "Got the work order you sent. I will have a look and get back to you." Or perhaps, "I saw your request. I'll try to get to it as soon as I can. Thanks." These will keep your relationships with co-workers moving smoothly.

當然要。只需以簡短 email 告知對方你已收到,就能讓發信者安心不少。你可回覆「Got the work order you sent. I will have a look and get back to you.」(收到你寄來的工作指示了,我會詳閱一番再回覆你。),或者「I saw your request. I'll try to get to it as soon as I can. Thanks.」(我看到你的請求信了。我會儘快著手處理,謝謝。)這些都能讓你和同事的關係更和睦。

如果是透過網站訂購頁面下單，通常會收到自動發出的確認 email。但如果是透過 email 訂購的話，卻不見得都能收到回覆。這種時候，就必須主動寄信確認。

Subject: Confirmation request: Order #3224

Jake,

[1]Could I confirm that you received my supply order last week? [2]I ordered via your website. [3]The order number is #3224. [4]One of the parts we ordered, the oven thermostat, is pretty urgent. [5]If you could check on that and get back to me, I would appreciate it.

Yosuke Tachihara
Tachihara Bakeries

翻譯

主旨：請確認：訂單 #3224

Jake：

我想確認一下您是否有收到我上週送出的訂單？我是透過貴公司網站訂購的。訂單編號為 #3224。在我們所訂購的零件中，有個烤箱恆溫器我們急需使用。如果您能確認一下並作回覆，我將萬分感激。

立原麵包店
立原洋介

重要片語 & 句型

■ **Could I confirm that you received my order?** 能否確認一下您是否有收到我的訂單？
■ **I ordered via your website.** 我是透過貴公司網站訂購的。
■ **The order number is ...** 訂單編號為～
■ **urgent** 緊急的
■ **If you could check on that and get back to me, I would appreciate it.**
　　如果您能確認一下並作回覆，我將萬分感激。

佳句便利貼

■ Could I have your order number please?
能否給我您的訂單編號？

■ My order confirmation number is 122863.
我的訂單編號是 122863。

■ Could I just make sure you are shipping our order on Wednesday?
能否確認一下我們的貨是否將於週三送出？

■ Could you check on that for me?
這部分可以幫我確認一下嗎？

■ One item you ordered is out of stock.
您所訂購的商品之一已無庫存。

■ I sent you a message through your website.
我透過貴公司網站寄了一封 email 給您。

 **FAQ 27** 敲定生意的關鍵

Question

What is the proper way of indicating the date? Wednesday, May 2, 20XX / Wed. May 2nd, 20XX / Wednesday, 2nd May, 20XX—there are several ways. We are often confused.

日期的正確寫法為何？ Wednesday, May 2, 20XX / Wed. May 2nd, 20XX / Wednesday, 2nd May, 20XX 等，寫法有好多種，我們常覺得很困擾。

Answer

Each English-speaking country uses a slightly different way. In the U.S. you can say "The meeting will be Wednesday, May 2nd at 2 p.m." or "The meeting will be on Wednesday, the 2nd of May at 2 p.m." In Britain, you might see "The meeting will be on 2 May, Wednesday, at 2 p.m."

每個英語系國家的日期寫法稍有一些不同。如果在美國，你可以說「The meeting will be Wednesday, May 2nd at 2 p.m.」，或者是「The meeting will be on Wednesday, the 2nd of May at 2 p.m.」；在英國則可能看到「The meeting will be on 2 May, Wednesday, at 2 p.m.」的寫法。

當訂購商品遲遲未送達時，就該馬上與對方聯繫、確認。如果遇到很緊急或對方一直沒回信的情況，那麼打電話詢問會比較保險。

Subject: Missing order: #14478

Dear Shipping Department,

[1]Nearly two weeks ago I put in an order (#14478) for two replacement conveyor belts. [2]I ordered them to be sent by express mail, but as of this morning they still have not arrived. [3]One of our belts snapped this morning and we are in desperate need of those replacements. [4]Could you please ship our order immediately? [5]Thank you.

Taro Hatoyama
Beltline Snacks

翻譯

主旨：訂購商品未送達：#14478

運送部：

大約 2 週前，我訂購了 2 個更換輸送帶（訂單編號 #14478）。訂購時指定以快遞送出，但是到今天上午為止都還沒送到。今天早上我們有一條輸送帶斷了，急需更換。能否請您立刻將我們訂購的商品送來？謝謝。

Beltline 點心食品公司
鳩山太郎

 重要片語 & 句型

■ **I put in an order for** ... 我訂購了～。

■ **as of this morning** 到今天上午為止

■ **We are in desperate need of** ... 我們急需～。

■ **Could you please ship our order immediately?** 能否請您立刻將我們訂購的商品送來？

佳句便利貼

■ We are in urgent need of color toner cartridges.
我們急需彩色碳粉匣。

■ We are desperate to get those parts.
我們非常需要那些零件。

■ Please have that delivered by courier.
請以快遞送來。

■ Please have the parts flown in from Sweden.
請將這些零件從瑞典空運過來。

 FAQ 28 敲定生意的關鍵

Question

What is a polite way to prompt a quick response? 如何有禮貌地要求對方馬上回信？

Answer

"I would appreciate it if you could get back to me as soon as possible." or alternatively, you could say "Please get back to me at your earliest convenience." Both of these will work, and you should always end with "Thank you."

建議你可以使用「I would appreciate it if you could get back to me as soon as possible.」（如果您能儘快回覆，我將萬分感激。）這種寫法，或者是「Please get back to me at your earliest convenience.」（請儘速和我聯絡。）這兩種表達方式都行得通，但要記得在最後加上「Thank you.」。

訂購大型商品或特製商品時，有時需由訂購者本人當面點收貨品。因此，如果能如下例那樣，具體告知交貨時間與運送方式等，就會顯得十分貼心。

Subject: Today's delivery

Claire,

[1]Thanks again for your order. [2]This is just a heads-up to let you know that your order is on the truck and should be delivered about 1 today. [3]The driver's name is Damien and, if he gets delayed, I gave him your number and asked him to call. [4]If I can be of further help, feel free to let me know. [5]We appreciate your business!

Hisako Fujita
Computer World

翻譯

主旨：今日出貨通知

Claire：

再次感謝您的訂購。本信的目的是要通知您注意，您所訂購的商品已經裝車出貨，大約今天下午 1 點左右會送達。運送司機的名字是 Damien，如果他會遲到，我有給他您的電話，請他務必與您聯繫。如果還有其他需要幫忙的，請告訴我一聲，不用客氣。感謝您的惠顧！

電腦世界公司
藤田久子

重要片語 & 句型

- **Thanks again for your order.** 再次感謝您的訂購。
- **This is just a heads-up to let you know that** ... 本信是為了通知您注意～。
- **If I can be of further help, feel free to let me know.**
  如果還有其他需要幫忙的，請告訴我一聲，不用客氣。
- **We appreciate your business!** 感謝您的惠顧！

佳句便利貼

■ Thanks for the heads-up!
感謝您的提醒！

■ We will ship you the remaining items once they are in stock.
待其餘商品一入庫，我們就會馬上送出。

■ Thank you for your business!
感謝您的惠顧！

 FAQ 29 敲定生意的關鍵

Question
Should we send any seasonal messages, say, for Christmas or Halloween? Is email enough instead of cards?

聖誕節或萬聖節時，是否該寄送祝賀節慶用的 email？email 是否能取代實體的卡片？

Answer
While more people are starting to send email greetings, unless it is part of another message, this seems unprofessional. Christmas cards are traditional and always look good. Since Chinese New Year is very important in Taiwan, a New Year card may make a positive impression as well, even to your foreign clients. But, those are generally the holidays when it would seem normal to send a card.

雖然有愈來愈多人開始寄送 email 賀卡，但是除非該 email 還包含了其他訊息，否則這種做法不很專業。寄送實體聖誕卡是一種傳統，也總是十分討喜。在台灣，農曆新年非常重要，所以實體的新年賀卡也能帶給對方好印象，即使是外國客戶亦然。不過在正式的商務往來中，一般也只有聖誕節和新年才會寄送卡片。

4

付款
Payment

請求對方付款時,有時也會將請款單以附件的方式一併寄出。由於請款金額等細節已明記於請款單中,故 email 裡只需打聲招呼、通知對方有附上請款單,並簡單註明購買商品與付款期限即可。

Subject: Payment request: Office window repairs

Mr. Jerry Smithers,

[1]Thank you for your business. [2]Enclosed please find the bill for the repairs to your office window. [3]If you could send your payment by August 31st, we would greatly appreciate it. [4]Thanks and we look forward to serving you again in the future.

Emi Suzuki
Archetype Windows

翻譯

主旨:請款:辦公室窗戶修繕

Jerry Smithers 先生:
感謝您的惠顧。隨信附上貴辦公室窗戶修繕之帳單明細。如果您能於 8 月 31 日前付款,我們將十分感激。謝謝,也期待未來能再次為您服務。

原型門窗公司
鈴木惠美

 重要片語 & 句型

■ **Thank you for your business.** 感謝您的惠顧。
■ **Enclosed please find ...** 隨信附上〜。
■ **If you could send your payment by ..., we would greatly appreciate it.**
　如果您能於〜前付款,我們將十分感激。
■ **We look forward to serving you again in the future.** 我們很期待未來能再次為您服務。

佳句便利貼

■ Attached to this memo is an invoice.
隨本信附上發票。

■ Here is the payment request for services rendered.
這是服務費請款單。

■ Your bill totals $32.48.
您的帳單總金額為 32.48 美金。

■ Please forward payment by the due date on your bill.
請在您帳單上所記載之截止日前完成付款。

■ Please pay immediately.
請立即付款。

■ Along with this message I have attached a coupon for your next order.
我已隨此信附上一張優惠券，可供您下次訂購時使用。

■ We hope to serve you again soon.
希望很快能再次為您服務。

FAQ 30　敲定生意的關鍵

Question

How can I be polite but request payment immediately?

如何才能有禮貌地請對方立刻付款？

Answer

Getting someone to pay is often difficult. One way to request immediate payment is to say "Please pay upon receiving this invoice." This is a very polite and formal way to request payment immediately.

要別人掏錢付款總是不太容易。有種要求對方立刻付款的說法是：「Please pay upon receiving this invoice.」（請在收到此發票後立即付款。）這是想請對方即刻付款時，相當禮貌且正式的表達法。

31 確認付款方式
Asking about the Method of Payment

付款方式會因國家、地區、業種不同而有所差異。透過 email 往來的企業主要都以信用卡付帳。VISA 和 Master 在大部分地區都可以使用，如果是其他種類的信用卡，就需再做確認。

Subject: Payment question

Luke,

[1]I'm just checking to see what forms of payment you accept. [2]Do you take credit cards? [3]If so, which ones? [4]Can I do a bank transfer? [5]If none of those is available, let me know and I will stop by and pay in cash.

Thanks.

Tomoko Saito

翻譯

主旨：關於付款的問題

Luke：
我想確認一下貴公司接受哪些付款方式。你們接受信用卡嗎？如果接受，哪幾種卡可以？我是否可以用銀行轉帳？若以上這些方式都不行，請通知我一聲，我會親自過去以現金付款。

謝謝。
齋藤朋子

 重要片語 & 句型

■ **I'm just checking to see what forms of payment you accept.**
我想確認一下貴公司接受哪些付款方式。

■ **Do you take credit cards? If so, which ones?**
你們接受信用卡嗎？如果接受，哪幾種卡可以？（Note: 請將兩句一起記起來）

■ **Can I do a bank transfer?** 我是否可以用銀行轉帳？

■ **If none of those is available, let me know.** 若以上這些方式都不行，請通知我一聲。

■ **pay in cash** 以現金付款

■ Which methods of payment will you accept?
你們接受哪幾種付款方式？

■ Can I write a check?
我可以開支票嗎？

■ We are a cash-only business.
我們只接受現金付款。

■ We expect a cash payment by the close of business today.
我們希望能在今日營業時間結束前收到現金。

■ I will forward the money to your First Hawaiian bank account #31256-8 today.
我今天會將錢匯入您在 First Hawaiian 銀行的帳戶 #31256-8 中。

■ I will move the money into your account upon completion of the work.
工作一結案，我就會將錢匯入您的帳戶。

 FAQ 31 敲定生意的關鍵

Question
What is the proper way of asking if my email has been received?

該如何詢問對方是否收到 email ？

Answer
"Did you receive my earlier message about ...?" is the phrase you are looking for. This is useful in times where you sent something but did not receive a response.

就用「Did you receive my earlier message about ... ?」（您是否有收到我先前寄出有關～的 email ？）這樣的表達方式即可。當你寄出 email 卻沒收到回應時，這句話相當實用。

商品送達時，通常會同時收到發票（invoice）。發票就等於是送達商品的明細表兼請款單。但是如果該商品並非物品（goods），而是服務（services）的話，那就必須郵遞或以 email 寄送發票。

Subject: Invoice: Murphy case

Dear Mr. Sulley,

[1]Here is the invoice for the consulting services we provided on the Murphy case. [2]The total for 32 hours of work comes to $1,007.42. [3]Attached to this message is a breakdown of the costs. [4]If anything is out of the ordinary, please let us know. [5]Payment is due by the end of the business week.

Regards,

Ken Oishi
Oishi, Watanabe and Kato Law Partners

翻譯

主旨：發票：Murphy 案

Sulley 先生，您好：
這是敝公司所提供 Murphy 案諮詢服務的發票。總工時 32 小時，合計 1,007.42 美金。隨信附上費用明細。如果有任何問題，請通知我們。付款期限為週五前。

祝好
大石 ‧ 渡邊 ‧ 加藤律師事務所
大石健

 重要片語 & 句型

■ **Here is the invoice for ...** 這是～的發票。
■ **The total for** A **hours of work comes to** B. 總工時 A 小時，合計 B（金額）。
■ **Attached to this message is a breakdown of the costs.** 隨信附上費用明細。
■ **If anything is out of the ordinary, please let us know.** 如果有任何問題，請通知我們。
■ **Payment is due by ...** 付款期限為～。

佳句便利貼

■ Here is your bill.
這是您的帳單。

■ Your total is $21.07.
總金額為 21.07 美金。

■ With tax your total amounts to $127.14.
含稅後總金額為 127.14 美金。

■ This a summary of the costs.
這是費用摘要。

■ Please render payment by the close of business on Friday.
請在週五營業時間結束前完成付款。

■ I would like to see a cost-by-cost list of our expenses.
我想看一下費用明細表。

■ Please do the work now and bill me later.
請先開工，之後再向我請款。

 FAQ 32 敲定生意的關鍵

Question
When you notice your business client or someone seems to have forgotten an agreement or promise, not necessarily on purpose, how can I ask them to follow through?

當發現客戶或是某人，（不一定是故意）似乎忘了先前的協議或約定時，該如何要求他們依約行事？

Answer
We have a very nice indirect phrase for such situations. "Did you get the chance to ...?" as in "Did you get the chance to finish the Johnson contract like we talked about? If not, could you please take care of that and get back to me?" This form is meant to be indirect and is simply a reminder of something you talked about with the person before.

對於這種情況，有個很不錯的委婉句型可用，那就是「Did you get the chance to ... ?」（你是否有做～？）例如「Did you get the chance to finish the Johnson contract like we talked about? If not, could you please take care of that and get back to me?」（你是否有按照我們討論過的方式擬好 Johnson 合約了？如果還沒有，能否處理一下然後再回覆我？）這種說法，就是特意以間接方式提醒對方你和他已談過的事情。

請求重開帳單
Requesting a Reissued Bill

遺失發票或請款單時，必須馬上請對方重開一張。如果是 email 附件形式的發票，只要請對方再寄一次就行了，應該不會造成太大困擾。

Subject: Invoice request

Dear Ms. Pratt,

[1]I have been looking everywhere, but I seem to have misplaced the bill you sent over the other day. [2]Could I ask you to reissue the invoice? [3]Once I get it, I will be sure to pay it off promptly. [4]I'm sorry for asking you to do this again.

Ken-ichi Ishiyama

翻譯

主旨：請開帳單

Pratt 小姐，您好：

我到處都找過了，但是我似乎把您前幾天寄給我的帳單弄丟了。能否請您重開一張發票？我一收到就會立刻付款。真的很抱歉同一件事讓您再做一次。

石山健一

 重要片語 & 句型

■ **I seem to have misplaced the bill you sent over the other day.**
我似乎把您前幾天寄給我的帳單弄丟了。

■ **Could I ask you to reissue the invoice?** 能否請您重開一張發票？

■ **I will be sure to pay it off promptly.** 我一收到就會立刻付款。

■ **I'm sorry for asking you to do this again.** 很抱歉同一件事讓您再做一次。

佳句便利貼

■ Would you mind sending the invoice again?
能否再寄一次發票給我？

■ Could I trouble you to reissue the bill?
能否麻煩您重開一次帳單？

■ I'm sorry to ask you to repeat the process all over again.
很抱歉同一件事讓您重頭再做一遍。

FAQ 33　敲定生意的關鍵

Question

What should I say in an email message if someone stops mailing me suddenly and I don't know why?

如果某人突然不再來信，而我不知道原因時，該怎麼寫 email 詢問？

Answer

If someone stops emailing suddenly, there is usually a reason. Perhaps they are busy, or perhaps you angered them. In any event, you can send a quick message saying something like "I haven't heard from you for a while. You must be very busy. Could I check on …?" If they still don't answer, then you probably won't hear from them again.

如果某人突然不再來信，通常都是有原因的。有可能他們太忙，也有可能是在生你的氣。無論原因為何，你都可以寄一封內容類似這樣簡短的 email：「I haven't heard from you for a while. You must be very busy. Could I check on … ?」（很久沒收到您的信了，您一定很忙吧？是否能幫我確認一下～？）如果對方仍不回應，那你可能就再也無法與他們取得聯繫了。

已付款通知
Notifying of a Payment Made

如果是採現金或信用卡付款，就不需特別通知對方，但如果以匯款、轉帳至指定帳戶的方式付錢，對方可能因忙碌而難以頻繁地確認款項是否已匯入，所以主動通知對方是比較理想的。

Subject: Payment sent

Ryan,

[1]Just following up to let you know that I deposited the money for the office furniture into your corporate account this afternoon. [2]It should already appear in your account. [3]If there are any problems, let me know. [4]It was pleasure doing business with you.

Yukiko Sanada
Moana Surfboards

翻譯

主旨：款項已匯出

Ryan：
謹在此通知您，今天下午我已將辦公室家具的款項存入貴公司帳戶，現在這筆錢應該已經在貴公司帳戶裡。如果有任何問題，請通知我一聲。很榮幸能與您合作。

Moana 衝浪器材公司
真田由紀子

 重要片語 & 句型

- **Just following up to let you know that ...** 謹在此通知您～。
- **deposited the money for ... into your corporate account** 將～的款項存入貴公司帳戶
- **It should already appear in your account.** 這筆錢應該已經在貴公司帳戶裡。
- **If there are any problems, let me know.** 如果有任何問題，請通知我一聲。
- **It was pleasure doing business with you.** 很榮幸能與您合作。

佳句便利貼

■ The money you requested has been deposited into your account.
您要求的款項已存入您的帳戶。

■ I put money into your savings account last Thursday.
我上週四已將款項匯入您的存款帳戶中。

■ Just want to let you know I've made the payment.
謹通知您一聲我已付款。

■ Do you want the transfer to go to your business account or personal account?
您希望款項匯入公司帳戶還是私人帳戶？

■ I really enjoyed working with you.
與您合作真的相當愉快。

■ It would be a pleasure to work together again in the future.
很期待未來還有機會合作。

■ This is to inform you that your rent is overdue.
在此通知您，您的房租逾期未繳。

 FAQ 34 敲定生意的關鍵

Question

What message should I send if someone stops mailing all of a sudden and you think you may have offended them?

如果覺得自己可能冒犯了某人，而造成對方突然不再來信，那 email 該怎麼寫比較好？

Answer

A good answer in this situation might be "I'm afraid I may have said something that offended you. If so, I sincerely apologize." This may work. But it is better to try to avoid getting into this situation in the first place.

在這種情況下，你也許可以這樣說：「I'm afraid I may have said something that offended you. If so, I sincerely apologize.」（我想我可能說了什麼冒犯您的話。如果真是如此，我誠心地向您道歉。）這樣或許就能化解了。不過，最好一開始時就盡量避免陷入這種情況。

確認已收款
Acknowledging Receipt of Payment

確認帳款已匯入或轉帳成功後，就該馬上與對方聯繫。而收據除了以郵遞方式外，也可如下例這樣以 email 附件的形式寄送，再由對方列印出來留存。

Subject: Payment confirmation

Hi Jody,

[1]This is to let you know that we received your payment today. [2]Thank you for your prompt attention. [3]I have sent your receipt as an attachment. [4]Please print out a copy and keep it for your records. [5]Thanks again.

Jay Asano
GO Publishing

翻譯

主旨：確認款項已收到

嗨，Jody：
謹在此通知您，我們今天已收到您支付的款項。感謝您即時處理。我已將收據附上，請列印出來留存。再次感謝。

GO 出版公司
淺野 Jay

 重要片語 & 句型

■ **This is to let you know that we received your payment.**
謹在此通知您，我們已收到您支付的款項。
■ **Thank you for your prompt attention.** 感謝您即時處理。
■ **I have sent ... as an attachment.** 我已將～附上。
■ **Please print out a copy and keep it for your records.** 請列印出來留存。

佳句便利貼

■ Thank you for your transaction.
感謝您的交易。

■ Your payment came in today and we thank you.
您支付的款項今日已入帳,非常感謝。

■ Thank you for your quick response.
感謝您的迅速回應。

■ We are grateful to you for paying us promptly.
感謝您即時付款。

■ For accounting purposes, we recommended keeping a copy of the bill.
我們建議您留存一份帳單,以利會計管理。

 FAQ 35 敲定生意的關鍵

Question

Should I call myself Zhi-ming Chen or Chen Zhi-ming?

我應該自稱 Zhi-ming Chen(志明‧陳),還是 Chen Zhi-ming(陳志明)?

Answer

Since you are going to be writing in English, you should follow the custom of first name first and last name last. In other words, "Zhi-ming Chen." If you do it the other way (Chen Zhi-ming), people will think Zhi-ming is your last name and call you Mr. Zhi-ming. If you wish to keep the Chinese order of names, you can write "CHEN Zhi-ming" or "CHEN, Zhi-ming" or "Chen, Zhi-ming."

既然要用英文寫電子郵件,就應該遵循先名後姓的慣例。換句話說,應自稱 Zhi-ming Chen(志明‧陳)才對。若反過來寫(亦即 Chen Zhi-ming),外國人會以為 Zhi-ming 是你的姓,因而稱呼你 Mr. Zhi-ming。如果你希望保留原本的姓、名順序,則建議可寫成「CHEN Zhi-ming」、「CHEN, Zhi-ming」或「Chen, Zhi-ming」。

要求對方支付拖欠帳款的工作一點兒也不輕鬆。到底該用多強硬的態度來催促付款，實在很難拿捏。在以下範例中，明白寫出不排除採取法律手段，可說是態度相當強硬的 email。

Subject: Delinquent account notice

Mr. Bill Partridge,

[1]This is to notify you that your account as of July 31st has become delinquent to the amount of $432.76. [2]Please pay immediately or we will be forced to cut your service and/or take legal action. [3]We value you as a customer and would rather avoid taking such steps. [4]Your prompt attention in this matter would be appreciated.

Sincerely,

Sachiko Maruishi
Internet One

翻譯

主旨：帳款拖欠通知

Bill Patridge 先生：

在此通知您，截至 7 月 31 日為止，您的帳戶已拖欠 432.76 美元。請立即付款，否則我們將被迫停止對您的服務並（或）採取法律行動。我們視您為重要客戶，所以希望能避免走到這一步。如果您能儘速處理此事，我們將萬分感激。

Internet One 公司
丸石幸子
敬啟

 重要片語 & 句型

■ **This is to notify you that ...** 在此通知您～。
■ **become delinquent to the amount of ...** 已拖欠～（金額）

88

- **Please pay immediately.** 請立即付款。
- **We will be forced to cut your service and/or take legal action.**
  我們將被迫停止對您的服務並（或）採取法律行動。
- **We value you as a customer and would rather avoid taking such steps.**
  我們視您為重要客戶，所以希望能避免走到這一步。
- **Your prompt attention in this matter would be appreciated.**
  如果您能馬上處理此事，我們將萬分感激。

佳句便利貼

- Your account is in arrears.
  您的帳戶有拖欠款項。

- Your bill is past due.
  您的帳單逾期未付清。

- If you do not pay, we will have no choice but to call a collection agency.
  如果您不付款，我們將別無選擇，只能通知討債公司了。

- We would hope that such actions would not be necessary.
  我們希望不需要採取這類行動。

FAQ 36　敲定生意的關鍵

Question
Why do many English email messages use "We" instead of "I," even when the message is written by one person?

即使信是由某一個人所寫的，為何許多英文 email 仍用「We」自稱而不用「I」？

Answer
Using "We" as in "We thank you for your prompt payment." sounds more professional than "I," which sounds personal. The "We" also gives the impression that the whole company is speaking, rather than a single person. It is more natural in most cases to use "We."

用「We」自稱，例如「We thank you for your prompt payment.」（感謝您即時付款。），聽起來較專業，而用「I」則感覺較個人化。另外，「We」也能給人是代表整個公司發言而非代表單一個人的印象。

當你發現自己無法於期限內付款時，就應該馬上連絡對方，告知希望延後付款。最重要的是，務必在 email 中清楚寫明延遲理由與希望延至何時。記得必須謙恭有禮地提出你的請求。

Subject: Payment deferral request

Dear Walt,

[1]I'm really sorry for being late in paying for the drywall work you did for us. [2]My accountant is away from the office until next week due to a family emergency. [3]As a result, we are behind schedule on paying our contractors. [4]Would it be possible to defer payment until she returns on Monday?

Sincerely,

Naohiro Ishiyama
Ishiyama Builders

翻譯

主旨：請求延遲付款

Walt，您好：
真的很抱歉拖欠您砌牆工程的款項。由於我的會計師家中有急事，下週才能回來上班，以致我們對承包商付款的進度都落後了。能否延到她週一回來時再付款？

石山建設公司
石山直弘
敬啓

 重要片語 & 句型

■ **I'm really sorry for being late in** *Ving*. 我真的很抱歉延遲了～。
■ **be away from the office** 沒來上班；不在辦公室
■ **due to a family emergency** 因家中有急事

- **as a result** 因此（= consequently）
- **We are behind schedule on** *Ving*. 我們～的進度落後了。
- **Would it be possible to defer payment until ...?** 能否延到～時再付款？

 佳句便利貼

■ I apologize for the delay in making payment.
我為延遲付款一事道歉。

■ I'm sorry I didn't pay you on time.
很抱歉沒能準時付款給您。

■ Due to technical difficulties, we are currently unable to process online payments.
由於技術上有困難，我們目前無法進行線上付款作業。

■ We can't give you a refund because all sales are final.
我們無法退錢給您，因為所有特價商品一旦售出概不退換。

■ Can I pay you later?
我可以稍後付款嗎？

FAQ 37 敲定生意的關鍵

Question
How can I get directions to someone's office? 該如何詢問怎麼去對方的辦公室？

Answer
To get directions to someone's office, you can ask "Could you tell me how to get to your office?" This should get you the information you need. You can also check their website and find the map to get to their office.

想知道對方的辦公室怎麼去時，可以問「Could you tell me how to get to your office?」（能否請您告訴我該怎麼到貴辦公室？），這樣應該就能獲得你需要的資訊了。另外，當然也可上對方的網站，找找看有沒有辦公室地圖。

5

客訴處理‧道歉
Trouble Management / Apology

投訴產品問題
Complaining about a Product

想針對商品或服務寄送客訴信件時，必須具體指出問題所在，再要求換貨或退款。另外，若商品有破損等瑕疵，則最好以數位相機拍下存證，並將影像檔隨 email 寄出。

Subject: Order complaint

Dear Apparel Supply Company,

[1]I'm writing to complain about the uniform shirts I ordered from your company last week. [2]My order was for five shirts, each with my name Tomo above the pocket. [3]But what you sent was five shirts with Toma on the pocket. [4]Could you please remedy this situation? [5]Thank you.

Tomotaka Taniguchi

翻譯

主旨：商品瑕疵客訴

Apparel Supply 公司：
我寫這封信的目的是要針對我上週向貴公司訂購的制服襯衫提出客訴。我訂的是 5 件襯衫，且每件的口袋上方都必須有我的名字 Tomo。但是你們寄來的卻是 5 件口袋上方印著 Toma 字樣的襯衫。能否請你們針對這情況做點補救？謝謝。

谷口友隆

重要片語 & 句型

■ **I'm writing to complain about** ...　我寫這封信的目的是要針對～提出客訴。
■ **My order was** ...　我訂的是～。
■ **What you sent was** ...　你們寄來的卻是～。
■ **Could you please remedy this situation?**　能否請你們針對這情況做點補救？

佳句便利貼

■ I would like to file a complaint.
我要提出客訴。

■ I would like to express my dissatisfaction with your poor service.
我想針對貴公司的劣質服務表達不滿。

■ In my order, I requested one dozen 6-inch helicopter bolts.
訂單上我要的是 1 打 6 英吋直升機用螺栓。

■ I ordered two of each of the replacement motors.
我訂購了各 2 台更換用馬達。

■ What I received was incorrect.
我收到的商品不對。

■ Could you fix this mistake?
能否請你們更正這個錯誤？

■ I would appreciate it if you could correct my order.
如果你們能更正我訂的貨品，我將十分感激。

 **FAQ 38** 敲定生意的關鍵

Question

What if I ordered something but what I received was different from my order? How can I explain this to the company?

如果我訂的東西和送來的東西不一樣，該怎麼辦？我該如何向該公司投訴？

Answer

In this case you can say "There has been a mistake with my order. I ordered A, but I received B. Could you please fix my order?"

遇到這種情況你可以這樣表達：「There has been a mistake with my order. I ordered A, but I received B. Could you please fix my order?」（我訂的貨出錯了。我訂購的是 A，收到的卻是 B。能否請你們更正我訂的貨品？）

39 針對瑕疵品道歉
Apologizing for a Defective Product

對於顧客所提出的抱怨，基本上都應迅速做出回應。要是回覆慢了，不僅會讓對方更生氣，還可能演變成難以收拾的局面。若能妥善處理客訴，有時反而能贏得顧客的信賴。

Subject: Our sincerest apology

Dear Mr. Winters,

[1]Please allow me to apologize for the defective part you received from our company. [2]While we make every effort to insure that all parts are inspected, occasionally a defective one escapes our best efforts. [3]Bring in the part and we will be happy to replace it at no cost. [4]Please accept our apology. [5]Thank you.

Kenta Hosoki
The Parts Place

翻譯

主旨：我們誠摯地向您道歉

Winters 先生，您好：
請容我針對您從敝公司收到的瑕疵品致歉。雖然我們竭盡所能地確保所有零件都經過檢驗，但仍免不了百密一疏。請將該零件送來，我們會很樂意免費退換。請接受我們的歉意。謝謝您。

零件廣場公司
細木健太

 重要片語 & 句型

■ **Please allow me to apologize for ...** 請容我針對～致歉。
■ **We make every effort to V.** 我們竭盡所能地～。
■ **Occasionally a defective one escapes our best efforts.** 百密一疏
■ **We will be happy to replace it at no cost.** 我們會很樂意免費退換。
■ **Please accept our apology.** 請接受我們的歉意。

佳句便利貼

■ I was told that you got a malfunctioning part.
我聽說您收到了瑕疵品。

■ This is the broken part.
這就是壞掉的零件。

■ While we try our best, sometimes we fail.
雖然我們竭盡所能，但仍偶有失誤。

■ We will do our best to satisfy your business needs.
我們將盡最大努力來滿足您業務上的需求。

■ Please examine all of the parts before shipping them.
出貨前請先檢驗所有零件。

■ Our humble apologies for causing you trouble.
造成您的不便，我們深感抱歉。

■ Please forgive this mistake.
請原諒這次的錯誤。

 FAQ 39 敲定生意的關鍵

Question

Is it OK to sign a message with just the first initial and the last name?

email 署名是否可以只用名字開頭的大寫字母和姓氏？

Answer

Yes. It is acceptable to sign a message "Z.M. Chen," instead of "Zhi-ming Chen." Actually, it looks very formal from the view of the reader. This would be good for business people who prefer not to be called by their first name.

可以。英文電子郵件的署名可用「Z.M. Chen」（Z.M. 陳）這種寫法取代「Zhi-ming Chen」（志明·陳）。事實上，這種寫法從讀信者的角度來看是十分正式的。而對於不喜歡被稱呼名字的商務人士而言，這會是個不錯的辦法。

缺貨的理由不外乎暢銷商品生產不及，或倉管不善等，但是不論原因為何，被告知的顧客一定都會心生不滿。此時只能告訴對方確切的到貨日，或表示之後會再聯繫。

Subject: Your current order

Dear Ms. Langley,

[1]I regret to inform you that the office chair you ordered is currently out of stock. [2]We are not sure at this time when the next shipment will be received. [3]We can offer you the option of selecting a different chair or we will be happy to contact you when we get word on when the next shipment will arrive. [4]We apologize for this inconvenience.

Hinako Sagami
Priest Office Furniture

翻譯

主旨：關於您目前的訂單

Langley 小姐，您好：
很遺憾必須通知您，您所訂購的辦公椅目前缺貨中。我們現在還無法確定下次進貨會是何時。您可選擇換購不同的椅子，或者我們很樂意在知道下次到貨時間時與您聯絡。很抱歉造成您的不便。

Priest 辦公家具公司
佐上日名子

 重要片語 & 句型

■ **I regret to inform you that ... is currently out of stock.** 很遺憾必須通知您～目前缺貨中。

■ **We are not sure at this time when the next shipment will be received.**
我們現在還無法確定下次進貨會是何時。

■ **We will be happy to contact you when we get word on when the next shipment will arrive.** 我們很樂意在知道下次到貨時間時與您聯絡。

■ **We apologize for this inconvenience.** 我們很抱歉造成您的不便。

佳句便利貼

■ I'm afraid that product is no longer available.
那項商品恐怕已停售。

■ My apologies, someone just bought the last monitor we had of that type.
很抱歉，剛剛有人買走了最後一台該型螢幕。

■ We are out of that product at the moment.
那項商品目前已銷售一空。

■ Could I get a rain check until that product comes in?
能否提供我優先預購券，以便下次到貨時使用？

■ When we hear back on your order, we will let you know.
我們收到有關您所訂購商品的訊息時，就會通知您。

■ We can let you decide which loan is better for your business.
您可選擇對貴公司最有利的貸款方案。

■ We give you the choice of one large carpet or two smaller ones.
您可選擇要 1 張大毯子或 2 張小毯子。

 FAQ 40 敲定生意的關鍵

Question

What if someone complains to us that a part is defective, but we believe the customer broke the part? How can we explain this problem to the customer?

如果有客戶向我們投訴零件有瑕疵，但我們確信是對方自己弄壞時，該怎麼辦？我們該如何對顧客說明？

Answer

Arguing with a customer is always difficult, but it is sometimes necessary. In this case you should write, "After examining the part, we do not believe the part to be defective. Rather, we believe that the part was damaged by the buyer during installation. We are, therefore, unable to replace the part."

跟顧客講道理一向很難，但有時確有其必要。以此例來說，你最好這樣說：「After examining the part, we do not believe the part to be defective. Rather, we believe that the part was damaged by the buyer during installation. We are, therefore, unable to replace the part.」（在檢查零件後，我們不認為這是瑕疵品。我們認為該零件是在購買者安裝時弄壞的，因此我們無法換貨。）

41 投訴貨品運送延誤
Complaining about a Delivery Delay

貨品運送延誤算是一種重大過失。對於確認交貨期後才購買的一方來說，這形同違約。但此時只能將訂購編號或貨品追蹤編號告知賣方，要求對方趕快查出配送狀況。

Subject: Missing package #D48349374

Dear KEG Delivery,

[1]I have an express mail package that was guaranteed delivery by 3 p.m. yesterday. [2]As of the start of business this morning, we still do not have our package. [3]The tracking number is D48349374. [4]I would like to express my deep dissatisfaction with your service as these documents are vital to my business. [5]Please inform me of the status of my package.

Maiko Kaneko
Kaneko Fisheries, Ltd.

翻譯

主旨：未送達包裹 #D48349374

KEG 貨運：
我有件保證昨天下午 3 點前送達的快遞包裹，但是直至今早上班時仍未送到，貨品追蹤編號為 D48349374。由於這些文件對我的業務影響重大，因此我要對貴公司的服務表達深切不滿。請告知我包裹目前配送狀況。

金子水產公司
金子麻衣子

 重要片語 & 句型

■ **The tracking number is ...** 貨品追蹤編號為～。
■ **I would like to express my deep dissatisfaction with your service.**
　我要對貴公司的服務表達深切不滿。
■ **be vital to ...** 對～影響重大
■ **Please inform me of the status of ...** 請告知我～目前的狀況。

佳句便利貼

■ You guaranteed that you would have the work done today.
您保證過會在今天完成該項工作。

■ This product comes with no guarantees.
此商品未附保證書。

■ As of this morning, I haven't heard anything from the head office.
截至今日上午為止，我還沒收到總公司傳來的任何消息。

■ I'm very unhappy about the quality of your work.
我對貴公司的工作品質非常不滿意。

■ Your service is not up to par.
貴公司的服務根本不及格。

 FAQ 41 敲定生意的關鍵

Question
Should I put the important information right at the beginning of the message or near the end of the message?

到底該把重要訊息寫在 email 開頭處，還是在接近結尾的地方？

Answer
In English email it is important to get to the point of writing quickly. Therefore, it is normally better to get to point at the beginning of the message.

在英文商務郵件中，盡快寫出重點是很重要的。因此，把關鍵訊息寫在 email 開頭處通常較為理想。

關於商品遲交或配送延誤等問題，縱使錯在貨運業者，但從顧客的角度來看，訂購窗口就是商品販賣公司，因此賣方必須慎重道歉，並應以某種形式表達誠意。

Subject: Re: Package delay

Dear Mary Torres,

[1]We would like to express our deep regret that your package was not delivered on time. [2]We examined the tracking route and found the package got tied up in U.S. customs. [3]While this was beyond our control, we would like to offer you a voucher for free shipping on your next package. [4]We appreciate your business and sincerely regret that your package did not arrive on time.

Sincerely,

Kazuhiko Fujita
White Cat Express Mail

翻譯

主旨：回覆：包裹遞送延誤

Mary Torres，您好：
對於您的包裹未能準時送達一事，我們深表遺憾。在仔細調查配送路線後，我們發現該包裹卡在美國海關處。雖然這個部分超出了我們的掌控範圍，但是我們仍願意提供 1 張抵用券，您下次的包裹可免運送費。感謝您的惠顧，同時也為您的包裹未能準時送達一事表達誠摯歉意。

白貓宅急便
藤田和彥
敬啟

 重要片語 & 句型

■ **We would like to express our deep regret that your package was not delivered on time.** 對於您的包裹未能準時送達一事，我們深表遺憾。

- **on time** 準時（※ in time ＝及時）
- **got tied up in U.S. customs** 卡在美國海關處
- **While this was beyond our control, we would like to** *V.*
  雖然這個部分超出了我們的掌控範圍，但我們仍願意～。
- **We appreciate your business.** 感謝您的惠顧。
- **We sincerely regret that** ... 為～一事表達誠摯歉意。

佳句便利貼

- We sincerely apologize for neglecting to return your call.
  對於忘了回電給您一事，我們深感抱歉。
- It seems that there is a problem somewhere in our supply chain.
  看來我們供應鏈中的某個環節是有問題的。
- The delivery flight has been delayed due to bad weather.
  配送的貨機因天候問題而延遲了。
- Though weather delays are not in our control, we do regret the inconvenience it causes.
  雖然天候導致的延遲並非我們所能掌控，但是因此造成不便，我們仍深感遺憾。
- Here is a coupon for our new hamburger.
  這是本店新漢堡的折價券。
- Please accept this gift certificate for our new spa.
  請接受敝公司這張新 SPA 的禮券。
- This is a prepaid card for any of our services.
  這是一張適用於敝公司所有服務的預付卡。

 FAQ 42 敲定生意的關鍵

Question
What should we put in the subject line, our name or what we are writing about?

email 主旨欄該寫上自己的姓名，還是信件內容主題？

Answer
This is a difficult one. If you are contacting someone for the first time, in order to avoid the spam filter you might try both as in "Business Proposal from KY Entertainment" or "Need Information about your products: Tomohisa Tsuruta." After you have made contact, then just what the email relates to should be fine.

這確實是個難題。如果是第一次與某人聯繫，為了避免被當成垃圾郵件，你可以兩種寫法都試試看，如「Business Proposal from KY Entertainment」（KY 娛樂公司的業務提案），或者是「Need Information about your products: Tomohisa Tsuruta.」（請提供貴公司商品資訊：鶴田智久）。而一旦與對方取得聯繫後，email 主旨只寫內容主題即可。

在餐廳遇上服務不周的問題時，可以當場抱怨。但是如果錯過了表達時機，或於日後冷靜下來才寄 email 投訴的話，則可參考下例。

Subject: Poor service complaint

To the manager of Andre's Steak House,

[1]I'm writing to complain about the service at your restaurant this evening. [2]I reserved a table weeks ago for a business dinner with a client and, when we arrived, our reservation has been lost causing a 45-minute wait for a table! [3]In addition, our server, Jason, was rude and brought our meals at different times. [4]At no point did we feel welcome in the restaurant. [5]We will not be eating there again.

T. Takeda
Tanaka Financial

翻譯

主旨：客訴劣質服務

致 Andre's 牛排館經理：
我寫此信的目的是要投訴貴餐廳今晚的服務。我在數週前訂位要與一位客戶共進晚餐談生意，但是當我們到達餐廳時，卻發現訂位被取消了，我們足足等了 45 分鐘才有位子可坐！此外，我們的服務生 Jason 不僅無禮，並且未能幫我們兩人同時上菜，我們感覺一點都不受餐廳的重視。我們再也不會到你們的餐廳用餐了。

田中財管
竹田 T.

 重要片語 & 句型

■ **I'm writing to complain about** ... 我寫此信的目的是要投訴～。
■ **in addition** 此外（= besides / moreover）
■ **At no point did we feel welcome.** 我們感覺一點都不受重視。

佳句便利貼

- Let me vent for a moment.
  讓我發洩一下不滿。

- All complaints should be directed to the vendor.
  所有客訴都應直接傳達給供應商。

- We will not be investing in your company.
  我們將不投資貴公司。

- There is no chance that we will continue our business relationship with TR Holdings.
  我們不可能繼續和 TR 控股公司維持業務關係。

 FAQ 43 敲定生意的關鍵

Question

Is it OK to set a deadline for a response? If so, how do we do it politely?

是否可以訂定回覆期限？如果可以，如何寫較有禮貌？

Answer

Yes, particularly if it is urgent. To do so politely, use this sentence: "If you could get back to me by Monday, I would greatly appreciate it." This is polite, and will normally get you a quick response.

可以，尤其是針對緊急事項。如果想禮貌地表達，請用這個句子：「If you could get back to me by Monday, I would greatly appreciate it.」（如果您能於週一前回覆，我將十分感激。）這樣的說法很客氣，並且通常都能迅速獲得回應。

因員工疏失造成顧客不滿時，其上司或負責人就必須代表公司替該員工賠罪。除了陳述抱歉的語句外，更需表示出努力防止再犯與設法解決問題的誠意。

Subject: Your complaint

Dear Mr. St. Clair,

[1]Thank you for your comments regarding our service. [2]I deeply apologize that you did not enjoy your experience at Hallelujah. [3]At our restaurant, we strive to offer every customer the best of Japanese cuisine. [4]Clearly this time we failed. [5]I have spoken to the employees about your message and please be assured that we will continue to strive to improve our service.

With gratitude,

Yoji Kawamura
Manager, Hallelujah

翻譯

主旨：關於您的投訴

St. Clair 先生，您好：
感謝您對本店服務所提出的建議。對於您在哈雷路亞餐廳的不愉快經驗，我感到十分抱歉。在本餐廳，我們盡力提供每位顧客最優質的日本料理。很顯然這次我們失敗了。我已向員工傳達了您的意見，請相信我們一定會繼續努力改善本店的服務品質。

萬分感激。
哈雷路亞餐廳 經理
川村洋次

 重要片語 & 句型

■ **Thank you for your comments regarding our service.** 感謝您對本店服務所提出的建議。
■ **I deeply apologize that you did not enjoy your experience at ...**
對於您在～的不愉快經驗，我感到十分抱歉。

■ **We strive to offer every customer the best of** ... 我們盡力提供每位顧客最優質的～。

■ **Clearly this time we failed.** 很顯然這次我們失敗了。

■ **Please be assured that we will continue to strive to improve our service.**
請相信我們一定會繼續努力改善本店的服務品質。

 佳句便利貼

■ All comments pertaining to our service will be directed to the appropriate department.
所有關於敝公司服務的意見都會傳達給相關部門。

■ Your comments about our service are greatly appreciated.
感謝您對於敝公司的服務所提出的意見。

■ We are surprised that our services did not match your expectations.
很訝異我們的服務未能符合您的期待。

■ Due to customer dissatisfaction, our new natto ice cream will be discontinued.
由於客戶反應不佳，因此我們的新商品納豆冰淇淋將不再販賣。

■ We work hard to bring you the finest foods from around the world.
我們竭力提供您來自世界各地的最佳美食。

■ Obviously, mistakes have been made.
顯然錯誤已經造成。

■ Without a doubt, we need to raise the level of service to our customers.
毫無疑問地，我們需要提升客戶服務品質。

FAQ 44 敲定生意的關鍵

Question
Is it worth complaining? Will they actually do anything?

投訴是否有用？對方是否真的會改善？

Answer
Absolutely. Complaining helps a business improve its service and most companies do read the comments from customers. Complaining is one of the most difficult forms of writing for non-English speakers. The longer you are a user of the language, the easier and more natural it will be to complain in English.

當然有。投訴能幫助企業改善其服務，而且大部分公司都會認真讀取客戶建議。投訴信對母語非英語的人來說，是最難寫的一種。不過英文用得愈久，用英文抱怨就會變得愈來愈輕鬆、自然。

6

公告・祝賀
Business Announcements /
Congratulations

在商品或服務即將漲價前，先通知客戶或一般客人會比較有誠意，也較能避免流失顧客。此時應清楚表示漲價幅度和金額並說明漲價原委，以展現出尋求對方諒解的態度。

Subject: Price increase and discount offer

To all our valued customers,

[1]As a result of the poor economy, it is our unfortunate position to announce a price increase of 3% on all of our products effective May 1st. [2]We deeply regret the need to raise our prices. [3]In an effort to soften the blow to our customers, until the end of April we will be offering a 5% discount to all orders. [4]From all of us at Brio Plastics, we thank you for your business.

E. Brio
President, Brio Plastics

翻譯

主旨：漲價通知與折價服務

給我們所有尊貴的顧客們：

由於經濟不景氣，很遺憾我們必須宣布，從 5 月 1 日起敝公司所有產品都將漲價 3%。我們真的很抱歉必須走上漲價一途，而為了盡力減緩對顧客們的衝擊，即日起到 4 月底止，訂購所有商品皆提供 5% 的折扣。Brio 塑膠公司全體員工感謝您的惠顧。

Brio 塑膠公司 總裁
E. Brio

 重要片語 & 句型

■ **as a result of ...** 由於～

■ **It is our unfortunate position to announce a price increase of ...% on all of our products effective ...** 很遺憾我們必須宣布，從～起敝公司所有產品都將漲價～ %。

■ **effective** 生效的;有效的

■ **We deeply regret the need to raise our prices.** 我們真的很抱歉必須走上漲價一途。

■ **in an effort to** V 盡力～（※ in an attempt to V＝試圖～）

■ **From all of us at ..., we thank you for your business.** ～全體員工感謝您的惠顧。

佳句便利貼

■ There was a price hike in gas prices this week.
本週汽油價格大漲。

■ We are implementing a price freeze for all of 20XX.
20XX 年全年度我們都將實施凍漲政策。

■ As we strive for success, we must overcome many challenges.
為了努力達成目標,我們必須克服許多挑戰。

■ As we try to increase profits, we can't forget about thanking our customers.
在試圖提高利潤的同時,我們也不能忘了感謝顧客。

FAQ 45　敲定生意的關鍵

Question
What is a P.S. and is it OK to use it in a formal business email message?

P.S. 是什麼意思?是否可用於正式的商務 email 中?

Answer
P.S. means postscript, something that follows the main text, usually following the signature (like P.S. I saw your manager Bob on the golf course today.) It can be used in business email messages if the person you are writing to is considered a friend. But in formal messages where the relationship is purely business, it is best not to use P.S. messages after the text.

P.S. 這個縮寫是指 postscript（附筆）, 接於主要內容之後,通常加在署名下方。例如:「P.S. I saw your manager Bob on the golf course today.」（附帶一提: 我今天在高爾夫球場看到貴公司經理 Bob。）若發信對象算得上是朋友,那麼 P.S. 也可寫在商務 email 中。不過在一般 email 裡,當雙方純屬商務關係時, 就最好別在內文後加上 P.S.。

46 新產品發表
Announcing a New Product

對店鋪或個人發出新產品、新服務通知時，多半是透過型錄或實體信件來處理，但是以 email 宣布的方式正迅速普及。此類信件以清楚易懂的方式，簡潔地將賣點確實傳達出來為佳。

Subject: New Product: Release

Greetings!

[1]We at Soma Pharmaceuticals are pleased to announce the release of our new headache Release. [2]In recent trials, our product was shown to be better in tests than our competitors' products. [3]Release works in as little as five minutes. [4]Pick up Release today at your local pharmacy or contact your Soma representative for more details.

J. Horiuchi
Marketing Director, Some Pharmaceuticals

翻譯

主旨：新產品：「Release」

您好！

Soma 製藥很高興在此宣布，我們的新頭痛藥「Release」開始發售。在最近的實驗中證實，我們的產品比其他競爭者的產品更有效。「Release」在服用後短短 5 分鐘內便能產生效果。今天就請到您附近的藥局購買「Release」，或與您的 Soma 銷售人員聯繫，以取得更多資訊。

Soma 製藥 行銷總監
堀內 J.

重要片語 & 句型

■ **We are pleased to announce the release of** ... 很高興在此宣布，～開始發售。
■ **pick up** ... 購買～（口語的說法）
■ **at your local pharmacy** 到您附近的藥局
■ **Contact** ... **for more details.** 與～聯繫，以取得更多資訊。

佳句便利貼

■ Now announcing our newest tablet computer, Stanza.
在此為您介紹我們最新的平板電腦——Stanza。

■ I have an announcement to make.
我有件事要宣布。

■ Our moisturizer is without a doubt better than more expensive cosmetics.
我們的保濕乳液肯定比其他更貴的化妝品還優秀。

■ Our new telephone has been proven to cause less missed calls in the office.
經證明，我們的新型電話能減少辦公室的未接來電。

■ In 30 minutes or less, you can drive away in a brand new Frontier Chevrolet.
只需 30 分鐘或更短的時間，您就能把全新的 Frontier 雪佛蘭轎車開回家。

■ Get yours today!
今天就訂購！

 FAQ 46 敲定生意的關鍵

Question
Should we reply to a message with the original message after our reply?

回覆 email 的時候，是否該把原信內容留在下方？

Answer
It doesn't matter professionally. But in busy offices it may help the person to remember what they contacted you about and may help you get a faster response.

留不留都不影響您的專業。不過在繁忙的職場上，如果留下原信內容，或許能幫助對方想起聯繫你的原因，因而較快獲得回應。

祝賀新事業
Congratulating on a New Business

收到企業客戶或個人寄來的新事業或新店開張等訊息時，就應該回覆一封祝賀 email。如果是關係親近的對象，一定會很高興收到您鼓勵的話語。

Subject: Congratulations!

John,

[1]Congratulations on the opening of The Harbor Bar and Grill! [2]We were pleased to learn of your new business venture. [3]The restaurant business can be difficult, but we believe you will thrive in this new business. [4]If we can be of any assistance, don't hesitate to give us a call.

Jun Fujioka
Fujioka Supply

翻譯

主旨：恭喜！

John：

恭賀「港灣燒烤酒吧」開店大吉！我們很高興聽到您要發展新事業。餐廳經營並不容易，但是我們相信您的新事業一定能夠很興旺。如果有什麼我們幫得上忙的，別客氣，請來電告知。

藤岡器材公司
藤岡純

 重要片語 & 句型

■ **Congratulations on the opening of** … 恭賀～開店大吉。
■ **We were pleased to learn of** … 我們很高興聽到～。
■ **We believe you will thrive in** … 我們相信您的～一定能夠很興旺。
■ **If we can be of any assistance, don't hesitate to give us a call.**
　　如果有什麼我們幫得上忙的，別客氣，請來電告知。

佳句便利貼

■ Welcome to our grand opening!
歡迎光臨我們的盛大開幕式！

■ Congratulations on the start of your new business!
恭喜您開啓新事業！

■ This business opportunity only comes along once in a lifetime.
這樣的商機一生只會遇上一次。

■ If you work hard, you will succeed as a business owner.
只要勤奮工作，您將成為成功的企業老闆。

■ This business idea is a sure hit.
這個創業構想肯定會造成轟動。

■ Don't think twice about getting help if you need it.
如果您需要幫忙，請別猶豫立刻尋求援手。

■ If you need anything, give me a call at 453-7128.
如果您有任何需要，請撥 453-7128 這支電話給我。

 FAQ 47 敲定生意的關鍵

Question

I keep getting emails I don't care about from a company. How can I ask to get off a mailing list?

某公司經常寄給我許多與我不相干的 email。我該如何要求對方將我從郵寄清單中移除？

Answer

"Please remove me from your mailing list."—just a simple message like this one will normally work. If this doesn't work, then most email software will allow you to block the address or automatically send it to your junk mailbox.

通常只需「Please remove me from your mailing list.」（請將我從貴公司的郵寄清單中移除。）這樣簡單一句就行了。如果這樣做無效，那麼你可利用多數 email 軟體都具備的封鎖功能來封鎖該郵件地址，或是將信件自動移到垃圾郵件匣中。

48 退休／調職通知
Announcing Someone's Retirement / Transfer

當聯繫窗口的負責人員有調職、職位異動、退休等狀況時,一般都需要事先通知相關企業或個人客戶。而如果該名員工任職期間很長,還可寫出其服務年資,藉此讚揚其功績。

Subject: Retirement of Tsutomu (Tom) Ejiri

To all our valued members,

[1]We would like to announce the retirement of our long-time salesperson Tsutomu (Tom) Ejiri. [2]Tom has been with our company for 37 years and has worked with many of you over the years. [3]We deeply appreciate his service to the company. [4]For those clients who worked directly with Tom, his accounts will be handled by Aki Chiba.

Yasutaka Oe
One Light Credit Union

翻譯

主旨:江尻勉(Tom)退休通知

致所有敬重的夥伴們:
我們在此宣布,本公司資深業務員江尻勉(Tom)即將退休。Tom 在本公司已經服務了 37 年之久,這些年來他與各位當中的許多人都共事過。我們非常感謝他對公司的付出。原先直接與Tom 聯繫的客戶請注意,他負責的業務都將由千葉亞紀接手處理。

一光信用合作社
大江康隆

 重要片語 & 句型

■ **We would like to announce the retirement of ...** 我們在此宣布,～即將退休。
■ **have been with** *one's* **company for ... years** 在～已服務～年之久
■ **over the years** 這些年來
■ **We deeply appreciate** *someone's* **service to the company.**
　我們非常感謝～對公司的付出。

佳句便利貼

- **Ben is retiring next week.**
  Ben 下週將退休。

- **After a long career, Brenda is leaving the company to enjoy retirement.**
  在工作多年後，Brenda 即將離開公司享受退休生活。

- **Shinji has worked with our company for a number of years.**
  真治在敝公司已服務多年。

- **Mari joined our company in 20XX as a clerk.**
  麻理於 20XX 年進入本公司擔任職員。

- **We are thankful for your continued support of our fundraising campaign.**
  非常感謝您持續支持我們的募款活動。

- **Kanako Haga will take care of your secretarial needs.**
  芳賀加奈子將負責您的祕書工作。

- **All information regarding the Hensley account should be turned over to our new director.**
  所有與 Hensley 帳戶相關的資訊，都應移交給我們的新總監。

 FAQ 48 敲定生意的關鍵

Question

How can I announce a change of my company's website URL to our customers?

如何通知客戶我們公司網站的網址有所更動？

Answer

"We moved. Please visit us at our new URL at http://www.alphaproducts.co.jp."—write this, everyone will understand.

「We moved. Please visit us at our new URL at http://www.alphaproducts.co.jp.」（我們的官方網站搬新家了，請至新網址 http://www.alphaproducts.co.jp。） 只要這樣寫大家就會懂了。

負責聯繫窗口的人員若知道有人事異動、部門轉換等情況時，應事先通知相關企業或個人客戶，否則如果造成 email 無法聯絡到人的狀況，可就麻煩了。

Subject: New VP of Securities

To all Value Fund shareholders,

[1]On January 1st, fund manager Takumi Asano will be promoted to Vice-President of Securities Trading at DKB Financial Limited. [2]Taking over the management functions of our Value Fund will be Kumiko Ishizu. [3]Rest assured, your assets remain in good hands.

DKB Financial Limited

翻譯

主旨：新證券部副總

致各位 Value Fund 的股東們：
從 1 月 1 日起，基金經理人淺野拓海將被晉升為 DKB 財務管理有限公司的證券交易副總。而 Value Fund 的管理工作將由石津久美子接手。請放心，您的資產依舊能獲得良好的管理與運用。

DKB 財務管理有限公司

 重要片語 & 句型

■ *A* **will be promoted to** *B.*　A 將被晉升為 B。
■ **vice-president of** ...　～的副總
■ **take over** ...　接手～
■ **Rest assured,** ...　請放心，～
■ **remain in good hands**　仍能獲得妥善保管、處理等

佳句便利貼

■ Maki Evans will become the new head of operations at Morton Solvents.
Maki Evans 將成為 Morton 溶劑公司新的營運負責人。

■ Minoru Imahori will be taking charge of personnel from next week on.
今堀實下週起將擔任人事主任。

■ Assuming responsibility for customer complaints will be Michiko Azuma.
東美智子將負責客訴處理。

■ You can be certain that everything is under control.
請您安心，一切都在掌控之下。

■ Rest easy, your funds are protected with Charter Bank of Carbondale.
請放心，您的基金受到渣打銀行卡本代爾分行的保障。

■ Your money is safe with us.
您的資金在我們的保管下非常安全。

 FAQ 49 敲定生意的關鍵

Question

What things should we not normally notify other companies about by email?

有哪些事通常不會透過 email 來通知其他公司？

Answer

Like most countries, things that might make the company look bad (resignations, firings, etc.) should be avoided. Celebratory news related to someone that company might be connected with is OK, so long as it is related to business.

和大部分國家一樣，會影響公司形象的事（如辭職、解雇等）都該避免。而與公司有關的個人喜事，只要是與公務相關，就都沒問題。

公告新地址
Announcing a New Address

辦公室搬遷時，一定要清楚寫明新地址。而能將附近的交通設施、周遭環境，以及搬遷對業務上可能造成的影響與變化等都附帶一提的話，會更為理想。

Subject: We are moving!

To our loyal customers,

[1]After 52 years at our present location on Dole Street, Barney's Bed and Appliance is moving. [2]Catch us at new location at the corner of Kuhio Boulevard and 5th Avenue. [3]At our new location you will find 32,000 square feet of the highest quality furniture and bedding, always at a reasonable price. [4]Doors open at 10 a.m. on August 7th.

Your friends at Barney's Bed and Appliance

翻譯

主旨：我們搬家了！

致忠實的顧客們：

位於 Dole 街現址已有 52 年歷史的 Barney's 寢具公司即將搬遷。請至位於 Kuhio 大道和第 5 街交叉口的新店址參觀。我們的新店佔地 32,000 平方英呎，您將看到最高品質的家具與寢具，我們將以一貫的合理價格販售。新店將於 8 月 7 日上午 10 點開幕。

您居家的好朋友——Barney's 寢具公司 敬啟

 重要片語 & 句型

- **After ... years at our present location, ... is moving.**
  位於現址已有～年歷史的～即將搬遷。（※ move in ＝搬入；move out ＝搬出）
- **Catch us at new location at ...** 請至位於～的新店址參觀。
- **at a reasonable price** 以合理的價格
- **Doors open at ...** 於～開門營業。（※ Doors close at ... ＝於～關門；～營業時間結束）

佳句便利貼

- We have been at this site since 20XX.
  我們從 20XX 年起就在此營業。

- The office moved to its current location last month.
  本辦公室上個月才搬到目前位置。

- You can find us at www.gocon.co.jp.
  本公司的網址是：www.gocon.co.jp。

- We will be available for your questions from 10-3 on Saturday.
  週六我們的諮詢時間為上午 10 點到下午 3 點。

- Business starts at 9 on weekdays.
  平日營業時間從 9 點開始。

- We will be open for business on the 7th of February.
  我們將於 2 月 7 日開始營業。

 **FAQ 50** 敲定生意的關鍵

Question
Is there a good way in English to say " 託您的福 "?

「託您的福」這句話是否有合適的英文說法？

Answer
Yes. A rough translation is "because of your support" or "due to your support."

有。中文的「託您的福」可大致翻譯成「because of your support」或「due to your support」。

放假或暫停營業等訊息可利用 email 通知大家。和實體信件、電話等不同，同一封 email 可一次寄給多個對象，相當方便。

Subject: Holiday notice

To all our loyal customers,

[1]We will be closed for a holiday from March 13th-31st. [2]All orders received by midnight on March 10th will be processed ahead of the vacation. [3]Those received after March 10th will be processed beginning April 1st. [4]We appreciate your understanding.

翻譯

主旨：假日休業通知

致所有的忠實顧客們：

從 3 月 13 至 31 日為止，我們將因放假而暫停營業。所有在 3 月 10 日午夜前收到的訂單都將於假期前處理完畢。在 3 月 11 日之後收到的訂單則會於 4 月 1 日開始處理。感謝您的體諒。

 重要片語 & 句型

- **We will be closed for a holiday from ...** 我們將因放假而從～起暫停營業。
- **All orders received by ... will be processed ahead of the vacation.**
  所有在～前收到的訂單都將於假期前處理完畢。
- **Those received after ... will be processed beginning ...**
  在～之後收到的訂單則會於～開始處理。
- **We appreciate your understanding.** 感謝您的體諒。

佳句便利貼

■ Our offices will close on December 25th for the Christmas holiday.
12 月 25 日本辦公室將因聖誕節假日而關閉一天。

■ We will be away from the office from the 12th of November until the 20th.
從 11 月 12 日到 20 日為止，我們都不在辦公室。

■ I will be out of town next week.
我下週要出差。

■ Call us by the end of the day and we will set up an appointment.
請於今天內來電，我們好敲定一個會面時間。

 FAQ 51 敲定生意的關鍵

Question

Is there a simple phrase to politely tell someone that they misunderstood my meaning in a previous message?

有沒有什麼簡單的說法可以禮貌地告訴對方他們誤會了我上一封 email 的意思？

Answer

Yes. "There seems to be a misunderstanding. What I meant to say was ..." This very politely points out the mistake and gives the writer a chance to give the true meaning.

有。例如：「There seems to be a misunderstanding. What I meant to say was ...」（我們之間似乎有些誤會。我想說的是～。）這個句型相當有禮貌地指出了問題，並讓寫信者有機會釐清真正想表達的意思。

新任職者到任通知
Announcing a New Hire

新的經營幹部到任對客戶、股東、顧客來說都是很重要的訊息。故在這類 email 中，多半會附上新任職者的簡歷，甚至是本人所寫的寒暄語句等。

Subject: Press Release

---

[1]For immediate release

[2]Three Marks Industries is pleased to announce Shin-ichi Saito has been selected our new company president, replacing retiring president Mayumi Tanabe. [3]Mr. Saito was for the past 13 years president of Inventive Solutions and will bring a wealth of knowledge and experience to our company.

翻譯

主旨：新聞稿

即刻發布

三標工業很高興宣布，齋藤新一已獲選為本公司新總裁，接任即將退休的田邊真由美總裁。過去 13 年間，齋藤先生是創意方案公司的總裁，而他將為本公司帶來豐富的知識與經驗。

 重要片語 & 句型

■ **be pleased to announce ...** 很高興宣布～

■ **replace ...** 接替～

■ **for the past ... years** 過去～年間

■ **... will bring a wealth of knowledge and experience to our company.**
～將為本公司帶來豐富的知識與經驗。

佳句便利貼

- Did you know Mr. Fuji was chosen for our Board of Directors?
  你知道富士先生獲選為我們的董事嗎？

- I heard Jason Seals was promoted to CEO.
  我聽說 Jason Seals 已晉升為執行長。

- Anna will be taking over the duties for Mary, who is away on maternity leave.
  在 Mary 請產假期間，Anna 將承接她的工作。

- Tim spent the last two years working at Johnson Industries.
  Tim 前 2 年任職於 Johnson 工業。

- From 2002-20XX Meg consulted for a number of Fortune 500 companies.
  2002 至 20XX 年間，Meg 替許多《財星》500 大企業提供諮商服務。

- Beth's significant résumé assures her of good job prospects.
  Beth 豐富紮實的履歷確保了她良好的就業前景。

- The vast level of knowledge and experience sets us apart from other companies.
  豐富的知識與經驗使我們有別於其他公司。

 **FAQ 52** 敲定生意的關鍵

Question
What do we call " 聯絡方式 " in English?　　　英文的「聯絡方式」該怎麼說？

Answer

" 聯絡方式 " is "contact information" or "personal details" as in "Can I get your contact information?" or "Can I have your personal details?"

中文裡面的「聯絡方式」就是「contact information」或「personal details」，如「Can I get your contact information?」 或「Can I have your personal details?」（可以給我您的聯絡方式嗎？）

員工訃聞通常會寄發給經營幹部和（部分）其他員工，以及與其有直接業務關係的客戶窗口等。有時會將喪禮日期、會場地址及聯絡方式等做成附件，或直接寫在 email 內文中。

Subject: Funeral for Mr. Kenzaburo Miyagi

[1]Sunstrand Industries is sad to report the loss of our beloved president Mr. Kenzaburo Miyagi. [2]Mr. Miyagi was 71 years old. [3]A memorial service will be held on September 14th at 7 p.m. at Yasuragi Funeral Home. [4]Flowers and offerings to the family may be sent to the funeral home.

Sunstrand Industries

翻譯

主旨：宮城健三郎先生的喪禮

Sunstrand 工業在此悲痛地通知您我們所敬愛的宮城健三郎總裁辭世的消息。宮城先生享年 71 歲。追悼會將於 9 月 14 日晚上 7 時，在安寧殯儀館舉辦。給遺屬的鮮花與祭奠品等，請直接送至殯儀館。

Sunstrand 工業

 重要片語 & 句型

■ **be sad to report the loss of** ... 在此悲傷地通知您～辭世之消息
■ **loss** 死亡（= passing [較 death 委婉])
■ **beloved** 敬愛的
■ **A memorial service will be held on** ... 喪禮將於～（日期）舉辦。

佳句便利貼

- I'm sad to have to tell you about the death of my business partner.
  很難過必須通知您我生意夥伴的死訊。

- Did you hear about the passing of Mr. Misawa?
  您有沒有聽說三澤先生過世的消息？

- There will be a memorial service in honor of Mr. Baba tomorrow evening.
  明天晚上將舉行馬場先生的追悼會。

- The visitation will be from 6-8 p.m. at the funeral home.
  弔祭將從晚上 6 點至 8 點在殯儀館舉行。

- Gifts to the bereaved may be forwarded to the family.
  給遺屬的禮品請轉交家族成員。

 **FAQ 53** 敲定生意的關鍵

Question

I heard that many people don't send Christmas greetings anymore in the United States. Is this true? Why not?

我聽說在美國很多人都不寄聖誕卡了，這是真的嗎？為什麼？

Answer

The main reason is that not everyone celebrates the Christian holiday of Christmas, so many companies prefer to say "Happy Holidays". This includes Christmas, New Year's Day, Hanukkah and Kwanzaa.

主要原因是，並非每個人都慶祝聖誕節這個基督教節日，因此許多公司喜歡用「Happy Holidays」（佳節愉快）來相互祝福。這些節日包括了聖誕節、新年、光明節（在 12 月中持續 8 天的猶太教祭典）及寬扎節（從 12 月 26 日延續至 1 月 1 日，為非洲裔美國人的節慶）。

54 回應員工訃聞
Responding to News of an Employee's Death

收到企業客戶或個人傳來的訃聞時，務必馬上回傳哀悼文。撰寫時應慎選用字遣詞。

Subject: Our sincere condolences

To the employees of Teshio Industries,

[1]We were deeply saddened to learn of the loss of your valued president Mr. Yuji Watanabe. [2]Mr. Watanabe was a trusted partner and loyal friend to our firm for many years. [3]He will be missed.

With greatest respect and admiration,
The employees and staff of Bursa Halliflex

翻譯

主旨：獻上誠摯的慰問之意

致天鹽工業全體員工：
獲悉貴公司備受敬愛的總裁渡邊祐二先生過世之消息，我們深感悲痛。多年來，渡邊先生一直是位深獲信賴的夥伴，也是我們忠實的好朋友。我們會懷念他的。

謹致上最崇高敬意
Bursa Halliflex 全體員工 敬上

 重要片語 & 句型

■ **We were deeply saddened to learn of ...** 獲悉～過世之消息，我們深感悲痛。
■ **valued** 備受敬愛的
■ **trusted partner** 深獲信賴的夥伴
■ **loyal friend to ...** ～忠實的好朋友
■ **... will be missed.** 我們會懷念～；～將被懷念。
■ **with greatest respect and admiration** 致上最崇高敬意

128

佳句便利貼

■ It was with great regret that we learned the news of her death late.
我們非常遺憾這麼晚才得知她過世的消息。

■ Ron was one of the best colleagues I ever worked with.
Ron 是我所共事過最優秀的同事之一。

■ Ms. Honda was a valuable asset to our company.
本田小姐對本公司來說，是十分重要的人才。

■ Our office appreciates all the hard work she did for us.
我們十分感激她對本公司所付出的努力。

■ We wish to send our message signed "With highest regards,."
留言的末尾部分，我們希望能寫上「獻上最高敬意」這幾個字。

 FAQ 54 敲定生意的關鍵

Question

Should I be worried if my message is not perfect English?

我是否該擔心自己的 email 英文寫得不夠完美？

Answer

If you are trying to land a major business deal, then it is more important. But if you are the customer, then you can worry less. It is the company's job to help you even if your message has mistakes. The more important the business email is, the more you will want to pay attention to the details.

如果你正試圖爭取一筆大生意，那麼英文寫得好不好就比較重要。但如果你是顧客，那就比較不用擔心。即使你的 email 裡有錯，該公司仍有責任要幫助顧客。而愈是重要的商務 email，你就愈需注意細節。

確實掌握員工的職務異動、退休等情況並傳達給全體員工知道，是很重要的。有時也可將最後上班日和新單位，以及表揚其貢獻的文句等也都一併寫上。

Subject: Josh Stevens transfer

To all staff,

[1]This is to let you know that our auditor Josh Stevens will be transferring from our Tokyo office back to our branch office in Denver. [2]His last day on the job will be January 31st. [3]We would like to thank Josh for the contributions he made to our office during his five years here in Tokyo. [4]Please stop by his office and wish Josh a successful transition to his new position.

O. Arai
President, Excitement Unlimited

翻譯

主旨：Josh Stevens 職務異動

致全體員工：
在此通知各位，我們的稽核員 Josh Stevens 將從東京分公司轉調回美國丹佛的分公司。他在此的最後一天上班日為 1 月 31 日。我們要感謝 Josh 在東京的 5 年期間對本公司所做的貢獻。有空請順道到他的辦公室，祝他成功就任新職務。

Excitement Unlimited 總裁
荒井 O.

 重要片語 & 句型

■ **This is to let you know that ...** 在此通知各位～。
■ **auditor** 稽核員
■ **transfer from** *A* **to** *B* 從 A 轉調至 B
■ **His last day on the job will be ...** 他在此的最後一天上班日為～。

■ **We would like to thank** ... **for the contributions** ... **made to our office.**
我們要感謝～對公司所做的貢獻。

■ **wish** ... **a successful transition to** *one's* **new position** 祝～成功就任新職務

佳句便利貼

■ I have requested a move from accounting to sales.
我提出了從會計部轉至業務部的調職申請。

■ Your request for a change in assignment to Cancun has been rejected.
您申請調職至坎昆的要求已被駁回。

■ Gary will leave his job on the first of August.
Gary 將於 8 月 1 日離職。

■ Akiko's final day of work before maternity leave is this Friday.
亞紀子在產假前的最後上班日是本週五。

■ We hope that everything will go smoothly in your job hunt.
我們希望您在找工作時一切順利。

FAQ 55　敲定生意的關鍵

Question

How can I ask for a raise in pay?　　該如何要求加薪？

Answer

The best way is to stress what you bring to the company and what you have accomplished. There is no exact way to ask for a raise, but "I would like to discuss my current salary with you." is one way to open the dialogue between you and your boss.

最好的辦法就是強調你為公司帶來了什麼，又完成了什麼。要求加薪並沒有什麼固定的說法，不過「I would like to discuss my current salary with you.」（我想和您討論一下我目前的薪資。）算是句不錯的開場白，能開啟你和老闆間的對話。

56 回應調動、辭職及退休通知
Responding to a Notification of a Transfer, Resignation and Retirement

在職務異動或離職等的時候，往往會收到一些表達個人祝福或鼓勵之意的 email。此時必須一一回覆，而回信時務必將感激之情融入內容中。

Subject: Many thanks!

Dear Fernando,

[1]Thank you so much for your kind words of encouragement about my impending transfer back to Japan. [2]I've very much enjoyed getting to know you and it has been a great pleasure working with you those past three years. [3]I'm looking forward to getting back to my family in Japan and eating Japanese food again!

Taro

翻譯

主旨：非常感謝！

Fernando，您好：
非常感謝您因為我即將調回日本所說的鼓勵話語。很高興能認識您，也很榮幸在過去 3 年間能與您共事。我非常期待回到在日本的家人身邊，並再度享用日本美食！

太郎

 重要片語 & 句型

■ **Thank you so much for your kind words of encouragement about** ...
  非常感謝您因為～所說的鼓勵話語。
■ **impending** 即將發生的
■ **I've (very much) enjoyed getting to know you.** 很高興能認識您。
■ **It has been a great pleasure working with you.** 很榮幸能與您共事。
■ **I'm looking forward to getting back to my family.** 我非常期待回到家人身邊。

56

佳句便利貼

■ I appreciate what you said about me to the promotion committee.
很感激您在晉升評核委員會中替我美言了幾句。

■ I'm not looking forward to my upcoming return to England.
我並不期待即將返回英國。

■ Mr. Konno is being called back to Osaka today.
今野先生今天被召回大阪。

■ I can't wait to go back and see my kids.
我等不及要回去看孩子了。

 **FAQ 56** 敲定生意的關鍵

Question

How do I say " 請多指教／麻煩您了 " in English?　「請多指教」或「麻煩您了」的英文該怎麼說？

Answer

It depends on the context of situations where you wish to say it. If you say it when you meet someone for the first time, you can simply say "Nice to meet / see you." If you say it when you ask a favor, you simply put "Please" in your sentence. "Please" and "Thank you" are two very important expressions and using them frequently will help you build good relations with your co-workers or employees.

這要依說話當時的情境而定。如果是在第一次與人會面時，只要說「Nice to meet / see you.」就行了。而如果是在請求別人幫忙時，則在句子裡加上「please」即可。「Please」和「Thank you」是兩個非常重要的詞句，頻繁地使用它們有助於和同事或員工建立良好關係。

57 歡送會通知
Notifying a Farewell Party

替職務異動、離職或退休的人舉辦歡送會時，若要用 email 通知大家相關訊息，除了簡單提及將離開者的豐功偉業與今後計畫外，別忘了寫上舉辦地點、日期時間、費用等資訊。

Subject: Going-away party

Everyone,

[1]We will be having a going-away party for three members of our staff who will be leaving us: Porter Roberts, Shin-ichi Egashira, and Mako Matsuo. [2]Porter is retiring after 32 years on the job, Shin-ichi is returning to Kyoto to start his own firm, and Mako has accepted a new position as Financial Chief of Bank Three in Kobe. [3]The party will be held this Friday after work on the 5th floor. [4]Food and open bar will be provided.

Seiko Ito
Human Resources Director

翻譯

主旨：歡送會

各位：
我們將為 3 位即將離開我們的夥伴：Porter Roberts、江頭真一以及松尾真子舉行歡送會。Porter 在公司工作已滿 32 年而要退休，真一要回京都開設自己的公司，真子則是接受了神戶第三銀行的財務總監一職。歡送會將於本週五下班後在 5 樓舉行，現場會提供食物和開放吧檯。

人事部 總監
伊東聖子

 重要片語 & 句型

■ **We will be having a going-away party for** ... 我們將為～舉行歡送會。
■ **retire after** ... **years on the job** 工作滿～年而退休
■ **start** *one's* **own firm** 開設自己的公司
■ **accepted a new position as** ... 接受了新的～工作

134

■ **Food and open bar will be provided.** 現場會提供食物和開放吧檯。

■ **open bar** 開放吧檯（免費提供各式酒精與非酒精飲料，可無限暢飲。）

57

佳句便利貼

■ There will be a farewell party tomorrow night in the lounge.
明晚在交誼廳有個歡送會。

■ We are having a going-away celebration for retiring staff tonight.
我們今晚將為退休員工舉辦歡送會。

■ Olga founded her firm in 20XX.
Olga 於 20XX 年時創立了她的公司。

■ I'm going to start a new career as an IT consultant.
我將展開擔任資訊科技顧問的新職涯。

FAQ 57 敲定生意的關鍵

Question

How do I request that someone redo a job or improve further on something?

該怎麼要求某人重做某項工作，或是進一步改善某件事？

Answer

"Please redo this. It needs more work." is one way of asking someone to improve on the work they did. "This needs more work. Please keep working on it." is another way.

「Please redo this. It needs more work.」（請重做一遍，這樣是不夠的。）就是一種要求某人改善已經處理過的工作的說法。另外也可以說「This needs more work. Please keep working on it.」（這樣還不夠，請繼續努力。）

58 請病假
Taking a Sick Day

病假通知早點發出較好。如果遇到無法寄送 email 的情況，則最好打電話以口頭通知。請病假時只要簡單陳述症狀與看病、治療等預定行程，再加上致歉的詞句即可。

Subject: My absence

Dear Ms. Caruthers,

[1]I'm sorry to inform you that I won't be able to make it in to work today. [2]I'm under the weather with a high fever and I'm headed to the doctor. [3]In the event that you need anything, please feel free to contact me. [4]I apologize for any trouble this may cause.

Kazuhisa Uehara

翻譯

主旨：請假

Caruthers 小姐，您好：
很抱歉必須通知您，我今天沒辦法到公司上班。我身體不適發高燒，現在正要去看醫生。要是您需要什麼，請別客氣，就跟我聯絡。很抱歉造成您的困擾。

上原和久

 重要片語 & 句型

- **I'm sorry to inform you that I won't be able to make it in to work.**
  很抱歉必須通知您，我今天沒辦法到公司上班。
- **I'm under the weather with ...** 我身體不適～（症狀）。
- **be headed to ...** 正前往～（= be on *one's* way to ...）
- **In the event that you need anything, please feel free to contact me.**
  要是您需要什麼，請別客氣，就跟我聯絡。
- **I apologize for any trouble this may cause.** 很抱歉造成您的困擾。

58

佳句便利貼

■ I will be unable to come to the dinner tonight.
我今晚無法參加晚宴。

■ I'm afraid I'm unable to attend the meeting due to a family emergency.
因為家中有急事，我恐怕無法出席會議。

■ I'm not feeling very well.
我不太舒服。

■ I'm really sick.
我真的病了。

■ Should you need anything, call this number, 423-8761.
萬一您需要什麼東西，請電 423-8761。

 FAQ 58 敲定生意的關鍵

Question

In memos I sometimes see TBA. What does TBA mean?

在備忘錄中我有時會看到 TBA 這個縮寫。TBA 到底是什麼意思？

Answer

It means "To Be Announced." In cases where a meeting time or place is not yet decided, we can use this short form to tell that we will inform them later of the details.

TBA 指的是「To Be Announced.」（稍後公布）。在會議時間或地點尚未確定時，就可以用這個縮寫告訴大家稍後會通知細節。

7

日常業務
Daily Work

在今日，只要利用公司內部網路，便能靠 email 傳達各種聯絡事項。而像會議日期等重要資訊，一收到就該立即匯入行事曆。

Subject: Meeting Wednesday 10:30 a.m.

To all Pima staff,

[1]There will be a meeting this Wednesday at 10:30 a.m. in the staff room. [2]The topic of the meeting will be our current budget crisis. [3]All members, except those away on company business, are expected to attend. [4]If for some reason you cannot attend, please contact your supervisor.

Tetsuya Soga
V.P. of Finance, Pima

翻譯

主旨：週三上午 10 點 30 分開會

致 Pima 全體員工：
本週三上午 10 點 30 分我們將在員工休息室舉行會議。會議主題是關於我們目前所面臨之預算危機。除了出差的同仁外，所有員工都應出席。如果有特殊原因無法參加，請與您的主管聯繫。

Pima 公司 財務部副總
曾我哲也

重要片語 & 句型

■ **The topic of the meeting will be ...** 會議主題是～。
■ **except those away on company business** 除了出差的同仁外
■ **All members are expected to attend.** 所有員工都應出席。
■ **If for some reason you cannot attend, please contact ...**
　 如果有特殊原因無法參加，請與～聯繫。

佳句便利貼

■ The meeting is about upcoming outsourcing.
此會議討論的是即將進行之外包作業。

■ The agenda for the meeting is still undecided.
會議議題仍未定。

■ Several staff members are away on business trips.
有幾位員工出差去了。

■ Ask your superior if you have any questions.
如果有問題，請洽詢您的主管。

 FAQ 59 敲定生意的關鍵

Question

How do I ask someone for a favor?　　　如何請別人幫忙？

Answer

There are several ways to ask in person or by email. "Could you do me a favor?" "Would you mind doing me a favor?" and the less formal "I need a favor." All of these are used frequently among friends and colleagues.

欲當面或透過 email 請人幫忙時，有幾種表達方式可用。例如：「Could you do me a favor?」、「Would you mind doing me a favor?」（能不能請您幫我個忙？）還有較不正式的「I need a favor.」（我需要人幫忙。）這些說法在朋友與同事間都很常用。

會議記錄過去都用印刷品的形式分發，如果下次會議開始時，無人對上次的會議記錄提出異議，就代表大家都同意。而現今的會議記錄則大多透過 email 發送，在期限內如果無人提出異議，便視為已同意。

Subject: Meeting minutes

To all staff,

[1]Attached to this message are the minutes from last week's meeting. [2]Please review them, particularly if you could not attend. [3]If anything is incorrect, please inform me by the 12th. [4]Otherwise, the minutes will be assumed to be correct and binding.

Mina Yamada
Secretary to the President

翻譯

主旨：會議記錄

致全體員工：
附件為上週會議的會議記錄。請過目，尤其是未能參加會議的同仁們更應詳閱。如果發現任何錯誤，請於 12 日前通知我，否則此份會議記錄將被視為正確且具有效力。

總裁祕書
山田美奈

 重要片語 & 句型

■ **Attached to this message is [are]** ... 附件為～。
■ **minutes from** ... ～的會議記錄
■ **If anything is incorrect, please inform me by** ... 如果發現任何錯誤，請於～前通知我。
■ **otherwise** 否則
■ **binding** 有約束力的；有效力的

■ Please examine the text of our new ads for any mistakes.
請檢查一下我們新廣告中的文案是否有任何錯誤。

■ Could you look over what I wrote and give me some feedback?
能否請您看一下我寫的內容,並給我一點意見?

■ This press release can be assumed to be correct.
此篇新聞稿可被認定為正確無誤。

■ Is this document binding?
此文件是否具有效力?

 FAQ 60 敲定生意的關鍵

Question

What's the difference between "borrow" and "lend"?

「borrow」(借入)和「lend」(借出)有何差異?

Answer

We borrow something from someone. We lend something to someone. We can say, "Could I borrow your pen?" or "Could you lend me your pen?" but not *"Could you borrow me your pen?" Chinese learners of English sometimes get confused with the two verbs probably because "借入" and "借出" in Chinese use the only one word—"借".

我們向別人借(borrow)東西,但是借(lend)東西給別人。我們會說「Could I borrow your pen?」(我可以跟你借支筆嗎?)或者是「Could you lend me your pen?」(你可以把筆借給我嗎?),但不說「*Could you borrow me your pen?」。台灣的英語學習者偶爾會把這兩個動詞搞混,這可能是因為中文的「借」兩種情況都適用。

61 腦力激盪會議通知
Announcing a Brainstorming Meeting

在正式會議之前，為了交換意見而舉行之非正式集會，就稱為腦力激盪會議。
在這種比較輕鬆的氣氛中，往往能激發出於正式會議中難以產生的獨特創意！

Subject: Brainstorming get-together

To all interested employees,

[1]There will be an informal brainstorming meeting to drum up ideas for our fall promotional campaign. [2]In line with our new relaxed work policy, management has decided that the meeting will be held off-property at the Maruta's Tap on 6th Street. [3]Attendance is not mandatory, but we would like the input of as many people as possible. [4]Thank you.

Kazuo Hosoki

翻譯

主旨：腦力激盪聚會

給所有有興趣的同仁們：
我們將舉辦一場非正式的腦力激盪會議，來為我們的秋季促銷活動募集新點子。本著我們新的輕鬆工作政策，管理階層已決定，此會議將在公司外第 6 街的 Maruta's Tap 舉行。此聚會並不強制出席，但是我們仍希望大家都能踴躍提供創意。謝謝。

細木和夫

 重要片語 & 句型

■ **informal brainstorming meeting** 非正式的腦力激盪會議
■ **drum up ideas** 募集創意
■ **in line with ...** 與～一致；本著～
■ **The meeting will be held off-property at ...** 此會議將在公司外的～舉行。
■ **Attendance is not mandatory.** 不強制出席；可自由參加。
■ **We would like the input of as many people as possible.**
　我們希望大家都能踴躍提供創意。

佳句便利貼

■ Can you come up with a new slogan for our dictionaries?
你能不能替我們的字典想出個新標語？

■ We would like to hear your views on the stock's performance.
我們想聽聽您對股市表現的看法。

■ All opinions on the new store design will be welcomed.
我們歡迎所有針對新店面設計的意見。

■ Let's try to drum up support for this idea among our co-workers.
讓我們試著在同事間爭取對此構想的支持。

■ Our partner company's views are in line with our views.
企業夥伴的看法與我們的一致。

■ The opinion of the boss matches our opinion.
老闆的意見與我們的契合。

■ Attendance is optional at the meeting.
此會議可自由出席。

 **FAQ 61** 敲定生意的關鍵

Question
How formal should messages be between employees?

員工之間的 email 應該要多正式？

Answer
Normally, most messages between employees are much less formal and more in a friendly conversational style than those between businesses. The one exception to this is email to one's boss or supervisor. Then the words should be chosen more carefully.

一般而言，大多數員工間的 email 不會像公司間往來的 email 那麼正式，而會採取輕鬆的對話風格。唯一的例外就是，寄給老闆或上司的 email。這類 email 的用字還是應該小心謹慎。

專案參與邀請
Requesting to Participate in a Project

為了讓企劃案的決策與執行更順利，通常會組成專案小組來負責。有時會將擔任相關工作的員工都召集起來，而大多數都採取事前通知合適人選的方式。

Subject: Project participation request

Tony,

[1]I'm setting up a project team for the new Larabie account and I was hoping I could get you to participate. [2]The work involves meetings three days a week for a month, then once a month after for five additional months. [3]If you are able to participate, your expertise would be a great help. [4]Please let me know if you can join us.

Kana Fujimori

翻譯

主旨：專案參與邀請

Tony：
我正在替新的 Larabie 客戶組織專案小組，而我希望你能參加。這份工作頭一個月每週需開 3 天會，然後接下來 5 個月則每月開 1 次會。如果你能加入，你的專業知識將成為莫大助益。請告訴我你是否能夠加入我們的行列。

藤森佳奈

 重要片語 & 句型

■ **I'm setting up a project team for ...** 我正在替～組織專案小組。
■ **I was hoping I could get you to participate.** 我希望你能參加。
■ **involve** 牽涉到～；包含～（= include）
■ **Your expertise would be a great help.** 你的專業知識將成為莫大助益。
■ **Please let me know if you can join us.** 請告訴我你是否能夠加入我們的行列。

■ I'm putting together a group to work on product designs.
我正在組織一個商品設計團隊。

■ We've formed a committee to discuss our labor dispute with the management.
我們已經組織了一個委員會針對勞資糾紛與管理階層進行溝通。

■ I'm looking for someone whose specialty is employee training.
我正在尋找員工訓練專家。

■ Your skills are exactly what this company has been looking for.
您所擁有的技能正是本公司一直在尋找的。

■ You will be responsible for human resource management.
您將負責人力資源管理。

 FAQ 62 敲定生意的關鍵

Question

Can I call my boss by his/her first name in an email message?

我是否可以在 email 裡直呼老闆的名字？

Answer

If you are on a first-name basis with your boss, then yes. However, if you're in a large company and your boss does not know who you are, then it is better to call him/her Mr. Smith or Ms. Roberts.

如果你和老闆之間具有可相互稱呼名字的交情，那當然沒問題。不過如果是在大公司裡，老闆根本就不知道你是誰的情況下，那還是稱呼 Mr. Smith 或 Ms. Roberts 之類的比較好。

請求對方提供資訊的 email 有各式各樣不同的寫法。在此就讓我們看一個發送錄用通知給求職者，請對方提供雇用手續所需資料及文件的 email 範例。

Subject: Your future employment

Deshawn,

[1]My name is Naoko Yoshida, and I'm the Human Resources chief. [2]Thank you for your application. [3]We are so happy that you will be coming to work for us. [4]In order to finish the paperwork, we will need further information about your employment history and your tax information. [5]Could you please stop by this office with your social security card and one more form of identification? [6]Thanks.

Naoko Yoshida, Human Resources Chief

翻譯

主旨：關於您今後的工作

Deshawn：
我是吉田直子，人力資源部的主任。感謝您前來應徵，我們很高興您將就職於本公司。為了完成必要的書面作業，我們需要有關您工作經歷以及納稅的進一步資料。能否請您帶著社會保險卡和另一份身分證明文件至敝辦公室一趟？謝謝。

人力資源部主任
吉田直子

 重要片語 & 句型

- **Thank you for your application.** 感謝您前來應徵。
- **We are so happy that you will be coming to work for us.**
  我們很高興您將就職於本公司。
- **We will need further information about ...** 我們需要～的進一步資訊。
- **employment history** 工作經歷
- **tax information** 納稅資料

■ **Could you please stop by this office with ...?** 能否請您帶著～至敝辦公室一趟？
■ **social security card** 社會保險卡

佳句便利貼

■ Please complete these documents and give them back to me.
請填妥這些文件後再寄回給我。

■ Please give your completed questionnaire to the secretary.
請將填好的問卷交給秘書。

■ Do you have any work experience?
您有任何工作經驗嗎？

■ What is your tax status?
您的納稅狀況如何？

■ Do you have any ID?
您有任何身分證明嗎？

■ Do you have anything to verify your identity?
您有任何可證明身分的東西嗎？

 FAQ 63 敲定生意的關鍵

Question

I am working for a company in the United States and people around me keep using my first name and I would like to ask them by email to stop calling me by my first name. How can I do so?

我目前在美國的一家公司工作，周遭的人都直接叫我的名字。我想寄一封 email 請大家別再直呼我的名字，該怎麼寫才好？

Answer

Don't. You're trying to impose your culture on another country's culture. In most English-speaking cultures, the use of the first name is common. By asking others there to follow your custom, you will alienate yourself from your co-workers. Just understand that it is part of the business culture of that country and get used to it.

千萬不要。你這樣等於是把自己的文化強加在他國的文化上。在大部分英語系國家中，直呼名字是很平常的事。要求別人遵守你的習慣，將會造成你和同事間的疏遠。應該要理解這是該國商業文化的一部分，並習慣它才好。

64 回應合作邀請
Responding to a Request for Cooperation

對於籌備活動等的請求協助或加入某專案之邀約，如果想回覆願意接受時，在字句之間傳達出喜悅與幹勁就非常重要。而拒絕時則務必力求鄭重有禮。

Subject: Re: Your participation on my sales team

Dear Alejandro,

[1]Thanks for your request for my participation on your sales team. [2]I appreciate your consideration. [3]It would be my pleasure to join the team. [4]I hope I can contribute immediately and that we can all work together to increase our year-on-year profits. [5]I look forward to working with you.

K. Nakajima

主旨：回覆：邀請您加入銷售團隊

Alejandro，您好：
謝謝您邀請我加入您的銷售團隊。非常感激您的賞識，我很樂意加入貴團隊。我很希望能立即有所貢獻，並和大家一起創造比去年同期更高的利潤。很期待與您共事。

中島 K.

重要片語 & 句型

- **Thanks for your request for ...** 謝謝您邀請我～。
- **I appreciate your consideration.** 非常感激您將我納入考慮；非常感激您的賞識。
- **It would be my pleasure to** V. 我很樂意～。
- **I look forward to working with you.** 很期待與您共事。

佳句便利貼

- **Thanks for asking for my help.**
  感謝您來尋求我的協助。

- **I'm grateful for your message asking for my assistance.**
  很感激您來信尋求我的援助。

- **Thanks for thinking of me.**
  謝謝您想到了我。

- **I'm happy to be considered by you for this important job.**
  很高興您考慮讓我擔任此重要工作。

- **I believe I can make an impact in this company immediately.**
  相信我能馬上替公司帶來影響力。

- **I can't wait to work with you to complete this business plan.**
  我等不及與您一同完成這個企劃案了。

- **It's a pleasure to work together to increase this company's revenues.**
  很榮幸能為增加公司收益而共同努力。

 **FAQ 64** 敲定生意的關鍵

Question

Is it OK to call my boss "Boss" when sending him/her a message?

寄 email 給我老闆時,是否可直接稱呼他 / 她「Boss」?

Answer

No. While we do have some phrases directed at the boss where we call him or her boss "Yes, boss." or "You're the boss.," usually the word "boss" is used when we are talking about the boss to someone else such as "My boss is really strict." We don't open an email message with boss (*Dear Boss,).

不行。雖然在某些慣用語中,確實會將老闆稱為「boss」,比方說「Yes, boss.」(是,老闆。) 或「You're the boss.」(您說了算。),但是「boss」這個字通常用在我們和別人談論自己的老闆時,例如「My boss is really strict.」(我老闆真的很嚴格。) 寫 email 時絕對不會用 boss 這個字來開頭(如 *Dear Boss,)。

能由值得信賴的人那兒獲得人才等的介紹，對業務工作而言可說是極大助力。
而請人幫忙介紹的 email 內容可事先寫好並儲存為範本，以便日後隨時取用。

Subject: Contacts request

Bob,

[1]I have been assigned to take over several accounts including Maxwell Finance. [2]Since you have had close contact with them for several years, I was wondering if you could introduce me to those in charge there. [3]Your contacts would be very helpful as we're hoping to build on the relationship between our companies. [4]Thanks.

Toshifumi Shibutani

翻譯

主旨：煩請幫忙聯繫

Bob：
我奉命接手了幾個客戶，其中包括 Maxwell 財務管理公司。由於您多年來與該公司一直有密切聯繫，所以我在想，能否請您將我介紹給該公司的相關負責人員？我們正希望建立雙方公司間的關係，因此您的介紹將會很有幫助。謝謝。

涉谷敏文

重要片語 & 句型

■ **I have been assigned to take over** ... 我奉命接手了～。
■ **have close contact with** ... 與～有密切聯繫
■ **I was wondering if you could introduce me to those in charge there.**
　我在想，能否請您將我介紹給該公司的相關負責人員？
■ **Your contacts would be very helpful.** 您的介紹將會很有幫助。

佳句便利貼

■ Could you introduce me to the people you know at Berry-Scott?
能否請您將我介紹給 Berry-Scott 公司裡您所認識的人？

■ We hope to increase our business ties to your company.
我們希望強化與貴公司之間的商務關係。

■ For our relationship to grow, there must be mutual respect and trust.
為了增進我們之間的關係，我們必須互敬互信。

 FAQ 65 敲定生意的關鍵

Question

I seem to not be receiving company memos by email. What can I write to make sure that I receive them?

公司內部的 email 備忘錄我好像都沒收到。我該怎麼寫以確保我會收到這些備忘錄？

Answer

"I don't think I'm receiving the company memos. Please make sure I'm on the mailing list. Thanks." Send that message to the person in charge of sending out the memos so they can make sure you are on the list.

在此建議你可以將「I don't think I'm receiving the company memos. Please make sure I'm on the mailing list. Thanks.」（我似乎沒收到公司內部的備忘錄，請確認我是否有被列在收件人清單中。謝謝。）這樣內容的 email 寄給負責寄送備忘錄的人，讓他們確保你的名字被列入收件人清單中。

獲得他人介紹而與目標對象取得聯繫後，應發一封表達感謝之意的 email 給介紹人。而在經過一段時間之後，針對介紹達成的成果，再次向介紹人送出感謝信也很重要。

Subject: Thank you!

Gunther,

[1]Thanks so much for introducing me to Mike Rockwell at Rockwell Appliance. [2]Thanks to your help, we managed to increase our business with them an additional 14%. [3]We couldn't have done it without you. [4]Thanks so much.

Masatoshi Komichi

翻譯

主旨：謝謝您！

Gunther：
非常謝謝您將我介紹給 Rockwell 家電的 Mike Rockwell。由於您的幫助，我們能成功地將與該公司的交易量提升 14%。若是沒有您，我們一定無法辦到。萬分感激。

小道正敏

 重要片語 & 句型

■ **Thanks so much for introducing me to ...** 非常謝謝您將我介紹給～。
■ **Thanks to your help, we managed to V.** 由於您的幫助，讓我們能成功地～。
■ **We couldn't have done it without you.** 若是沒有您，我們一定無法辦到。

佳句便利貼

■ Thanks for everything you did.
感謝您所做的一切。

■ We would have been lost without your help.
若是沒有您的協助，我們肯定已經失敗了。

 FAQ 66 敲定生意的關鍵

Question
What does ASAP mean?

ASAP 是什麼意思？

Answer

It means "as soon as possible." We use it when we want something done now or within a short amount of time. We can use it in conversation too. In speaking, it is pronounced by spelling it out A-S-A-P, or it can also be said as an acronym ASAP.

ASAP 就是「as soon as possible」（儘快）。當我們希望某件事能馬上或短時間內完成，就可用此縮寫。它也可用在對話中，而口說時的發音方式就如拼字般，唸成 A-S-A-P，或採取字頭語的發音方式，唸成 ASAP（[əˋsæp]）。

宣告公司新規定
Announcing a New Office Regulation

本例是有關實施或修改新的公司內部規定時所用的公告信。這類郵件不只是要喚起員工注意而已，有時還會明確寫出違反者可能遭解雇等處罰的嚴正聲明。

Subject: Prohibition of SNS sites on company time

To all employees,

[1]Effective immediately, the use of social networking sites such as Friendbook and Twizzer during company time is prohibited. [2]Employees found to be using these sites on company time will be reprimanded and, if it continues, will be fired. [3]We apologize for this policy but it has become necessary.

K. Tomioka
CEO, West Pacific Courier

翻譯

主旨：上班時間禁止使用社群網站

致全體員工：
在上班時間禁止使用 Friendbook 和 Twizzer 之類的社群網站，此規定即刻生效。員工在上班時間如果被發現使用這類網站，將給予申誡，而如果持續再犯，便將遭解雇。很抱歉出此下策，但目前確實有必要。

西太平洋快遞 總裁
富岡 K.

重要片語 & 句型

■ **effective immediately** 即刻生效
■ **be prohibited** 禁止～（= be banned; be forbidden）
■ **be reprimanded** 被訓斥
■ **be fired** 被解雇
■ **We apologize for this policy.** 很抱歉出此下策。

佳句便利貼

- Starting today, the company will no longer pay for employee parking.
  從今天起，公司不再支付員工的停車費。

- From this point forward, our company will be called Hourglass Inc.
  今後本公司將改名為 Hourglass 股份有限公司。

- No personal calls on company time, please.
  上班時間請不要講私人電話。

- This is company time.
  現在是上班時間。

- Employees will be warned if they are breaking company rules.
  員工如果違反公司規定，就會被警告。

- This is your final warning.
  這是最後一次警告。

FAQ 67　敲定生意的關鍵

Question

I received a message from someone but it came to me as weird characters (the message was unreadable). How can I tell someone that their email looks that way?

我收到一封 email，但內容都是亂碼（無法閱讀）。該怎麼告知對方此問題？

Answer

There is no fixed way to describe "weird characters" in English since we normally have no problem like that in email in English. The best translation of "weird characters" is "unreadable." It is also possible to say "The message was all garbled up." You might want to say, "I'm sorry, but your message came to me in an unreadable form. Could you try sending it again in a different format?" Another way is, "I'm sorry, but your message came all garbled up. Could you send it again in a different way?"

在英文裡，「亂碼」並無固定說法，因為英文 email 通常不會發生這種問題。「亂碼」最理想的英譯就是「unreadable」（無法閱讀）。你可以說「The message was all garbled up.」（信件內文都亂成一團。）或「I'm sorry, but your message came to me in an unreadable form. Could you try sending it again in a different format?」（很抱歉，您寄來的信都是亂碼。能否請您試著以不同格式重寄一次？）還有另一種表達方式則為「I'm sorry, but your message came all garbled up. Could you send it again in a different way?」（很抱歉，您寄來的信都是亂碼。能否請您以不同方式重寄一次？）

在 email 普及以前，相對較不重要的資訊通常就以口頭形式傳達，因此漏聽的情況不在少數。現在則是什麼資訊都能透過 email 傳送，資訊遺漏的情形大減，大家變成忙於區別重要和不那麼重要的訊息，以及資訊整理的工作。

Subject: Re: Bakersfield order

Diane,

[1]Here is all the information I have on the current Bakersfield order. [2]If you need past order history, please contact Lou in accounting. [3]If I can be of further assistance, get in touch with me again.

Tomoko

翻譯

主旨：回覆：Bakersfield 的訂單

Diane：
這些是我手上所有和 Bakersfield 公司目前訂單相關的資訊。如果您需要以往的訂購紀錄，請與會計部門的 Lou 聯繫。如果還有什麼我幫得上忙的，請再次跟我聯絡。

朋子

 重要片語 & 句型

■ **Here is all the information I have on ...** 這些是我手上所有和～相關的資訊。
■ **If you need ..., please contact ...** 如果您需要～，請與～聯繫。
■ **order history** 訂購記錄
■ **accounting** 會計部門
■ **If I can be of further assistance, get in touch with me again.**
　如果還有什麼我幫得上忙的，請再次跟我聯絡。

■ Do you have the records for purchases over the past six months?
您有過去 6 個月的購買記錄嗎?

■ Here is a detailed list of everything ordered for the last two years.
這是過去 2 年間所有訂購貨品的細目清單。

■ I have no other knowledge except the information I have already given you.
除了已給您的資訊外,其他事情我並不清楚。

■ Give me a call if you need anything.
如果有任何需要,請撥電話給我。

■ Ring me if you need help.
若需幫忙,請打電話給我。

FAQ 68　敲定生意的關鍵

Question

I'm sending some very important information that should not be known outside the employees I'm sending it to. How can I express this to them?

我要傳送不能讓收件員工以外的人知道的極重要資訊。我該如何表達?

Answer

For protected information we should use "Secret" or "Top Secret" or " For Your Eyes Only," as in "This information is top secret." These mean that the information should be guarded and not given to anyone else.

針對必須保密之重要資訊,可使用「Secret」(機密)、「Top Secret」(最高機密)、「For Your Eyes Only」(請勿讓別人看見)等說法,例如「This information is top secret.」(此資訊為最高機密。)這些說法都表示該資訊應受謹慎保護,不可洩漏給他人。

69 伺服器維護公告
Announcing Server Maintenance

對於 email 已成為最重要通訊方式的現代辦公室來說,因伺服器維護而造成的網路停擺狀況,可算得上是與停水、斷電等不相上下的災難,因此有必要提早通知大家。

Subject: Computer server maintenance

To all employees,

[1]Please be aware that our company servers will be down for maintenance from 9 p.m. Friday, the 7th to 6 a.m. on Saturday, the 8th. [2]While our main customer site will still be accessible, there will be no access to the server until the work is complete. [3]Your understanding and cooperation on this matter will be greatly appreciated.

Y. S.

翻譯

主旨:電腦伺服器維修通知

致全體員工:
請注意,本公司伺服器因實施維修工程,將從 7 號週五晚上 9 點起至 8 號週六上午 6 點止暫停運作。雖然我們主要的客戶用網站仍可登入,但是在維護工程結束前,該伺服器都無法使用。非常感謝您對此事的體諒與合作。

Y. S.

 重要片語 & 句型

■ **Please be aware that** ... 請注意～。
■ **be down for maintenance** 因維修工程而暫停運作
■ **customer site** 客戶用網站(非公司內部使用,而是對外公開的一般網站或網頁。)
■ **Your understanding and cooperation on this matter will be greatly appreciated.**
  非常感謝您對此事的體諒與合作。

■ Please take notice that offices will be closed for the Easter holiday.
請注意，復活節假期期間辦公室將會關閉。

■ Please recognize that we all must find ways to cut costs.
請體認我們都必須設法降低成本。

■ The company car is in the shop for repairs.
公務車正在車廠裡維修。

■ The photocopier requires servicing.
影印機需要維修了。

■ There is no entry to the warehouse after 10 p.m.
晚上 10 點以後禁止進入倉庫。

■ The company safe is preset so that there is no way in after midnight.
公司保險箱已預設為午夜後無法開啟的狀態。

■ We must stay at our desks until the work gets done.
我們必須留在辦公桌前，直到工作完成為止。

 FAQ 69 敲定生意的關鍵

Question

What's the best way to turn down a request from a co-worker?

拒絕同事的最佳方式為何？

Answer

It depends on the situation. It is OK to say "I'm afraid I can't." In situations where it is work-related, it is best to give a reason why you can't do something. If it is personal (like an invitation to drink after work), then it is OK to say "I'm busy tonight." with no explanation.

這要視情況而定。例如你可以說「I'm afraid I can't.」（我恐怕沒辦法。）如果是與工作相關的請求，最好能提出你無法配合的理由。但若為私人事務（比方說下班後去喝一杯之類的事），那就可回答「I'm busy tonight.」（我今晚很忙。），不需多作任何解釋。

當辦公室增添了新的影印機等設備時，以往多半靠員工們互相教導來學習使用方法，但是現在這種教學的角色也由 email 取代，負責人員不需反覆進行同樣的說明。

Subject: New copier

Memo

[1]A new Rybor 4299 Photocopier/Scanner has been added to the office for your use. [2]This new photocopier will replace our two outdated Cabon 12 series copiers and the two desktop scanners we had. [3]A demonstration of the functions of the copier will be at 10 a.m. [4]Please attend.

Tomo

翻譯

主旨：新的影印機

備忘錄

本辦公室添購了一台新的 Rybor 4299 影印兼掃描機供各位使用。這台新影印機將取代我們兩台老舊的 Cabon 12 系列影印機和兩台桌上型掃瞄器。新影印機的功能示範教學將於上午 10 時開始，煩請參加。

智

 重要片語 & 句型

- **photocopier** 影印機（= copy machine）
- **replace ...** 取代～（※ replace A with B ＝用 B 取代 A）
- **outdated** 舊式的；過時的（※ state-of-the-art ＝運用最新科技的）
- **Please attend.** 煩請參加。

■ A presentation of the many uses of our new hair gel will start shortly.
我們新髮膠產品的多種用途展示即將開始。

■ This camera has many capabilities other models simply don't have.
此相機具備多種其他機型所沒有的功能。

■ Each button on the main panel performs a different action.
主面板上的各個按鈕都具備不同功用。

■ This computer is obsolete.
這台電腦過時了。

■ This projector is an older model.
這台投影機屬於舊式機種。

 FAQ 70 敲定生意的關鍵

Question

I got an invitation to the company picnic. The invitation said it is potluck and BYOB. What do these mean?

我收到公司野餐活動的邀請，邀請信中說是 potluck 與 BYOB 的形式。「potluck」和「BYOB」是什麼意思？

Answer

"Potluck" means that each person who comes to the party or picnic should bring a dish (prepared food to share with everyone). "BYOB" means Bring Your Own Booze/Beer/Bottle, and means if you want to drink alcohol you should bring what you want to drink. Both of these expressions are very common.

「potluck」是指每個參加派對或野餐活動的人，都該帶一道菜來（準備食物與大家分享）。而「BYOB」則代表 Bring Your Own Booze/Beer/Bottle，意思就是——如果想喝酒請自備。這兩種表達方式都很常見。

本例為辦公室裝修時的通知信。信中簡單提示了施工期、具體位置,以及對業務方面可能產生的影響等資訊。而如果有需要,甚至可考慮將具體的施工內容和草圖也附上。

Subject: Office renovation announcement

All Staff,

[1]We are happy to announce that our offices will take on a new look. [2]We have decided to renovate the existing work area, the storage rooms, conference room and lounge. [3]We anticipate the work will be complete by January. [4]We hope to minimize disruption to business, so the work will be completes in stages. [5]We hope the new changes will increase employee morale and make a better working environment for all.

T. Hasuike

翻譯

主旨:辦公室裝修通知

各位同仁:

在此很高興宣布,我們的辦公室即將煥然一新。我們決定重新裝修原有的工作區、儲藏室、會議室與交誼廳,裝修工程預計於 1 月前完工。而為了將對業務之影響降至最低,工程會分段進行。希望新的改變能提升員工士氣,並為大家創造更好的工作環境。

蓮池 T.

 重要片語 & 句型

- **be happy to announce that** ... 很高興宣布~。
- **take on a new look** 煥然一新
- **renovate** 裝修
- **existing** 現有的;原有的
- **We anticipate the work will be complete by** ... 此工程預計於~前完工。

- **minimize disruption to business** ... 將對業務之影響降至最低
- **be complete in stages** 分段進行；分階段完成

 佳句便利貼

- We put a fresh face on the lobby by painting it over the weekend.
  我們利用週末全面粉刷大廳，為它換上了全新的面貌。

- This shopping center is closed for renovations.
  此購物中心目前因重新裝修而暫停營業。

- We expect the drawings to be submitted tonight.
  我們期待設計圖今晚送到。

FAQ 71　敲定生意的關鍵

Question

My co-worker sent me his work. But it is not my job and I wish to refuse. How can I do so?

我同事把他的工作傳給我，但那根本不是我該負責的事，所以我想拒絕。我該怎麼做？

Answer

"I'm sorry but this work is not my responsibility." This is polite but strong. Different people, however, will react in different ways to this statement. So, it should be used with a small bit of caution.

「I'm sorry but this work is not my responsibility.」（很抱歉，這不是我負責的工作。）這樣的說法就夠有禮貌又堅決。不過每個人對這種回覆的反應都不同，所以使用時還是小心謹慎些為妙。

8

出差
Business Trips

出差指派通知
Ordering Someone to Make a Business Trip

本例為指派員工出差的 email。和口頭指派的情況不同，用 email 便可寫入詳細資訊，還能儲存下來。信中應包含具體的出差日期、時間，以及負責人員的聯絡方式。

Subject: Australia expansion

Maria,

[1]We need you to fly to Sydney next week to meet with the division heads of our four Australian outlets and to go over the expansion plans for Australia. [2]The meetings will be held on August 14-16, so we need you flying by the 12th. [3]Sandy Davis will handle all of your travel arrangements, so please see her right away. [4]Thanks.

Hiroshi Sakai
Peach Pit Fashions

翻譯

主旨：澳洲的業務擴展

Maria：
我們需要妳下週飛到雪梨去見我們澳洲 4 家暢貨中心的地區負責人，並討論關於澳洲的業務擴展計畫。會議將於 8 月 14 到 16 日舉行，因此妳必須在 12 日前出發。Sandy Davis 將幫妳處理所有的出差準備工作，請立刻去見她。謝謝。

Peach Pit 流行服飾公司
酒井廣志

 重要片語 & 句型

■ **We need you to** *V.* 我們需要你～（溫和的命令）
■ **fly to** ... 搭機飛往～（= travel to ... by airplane）
■ **go over** ... 討論～（= discuss）
■ **handle all of** *one's* **travel arrangements** 幫～處理所有的出差準備工作
■ **right away** 立刻（= right now / at once / immediately）

佳句便利貼

■ What are your travel plans for the exhibition?
你的展覽會出差計畫為何？

■ What is your itinerary for the upcoming business trip to Taipei?
你即將出發的台北出差行程路線安排為何？

■ We want you to look over these plans and give us your opinion.
請看一下這些計畫，並且提供你的意見。

FAQ 72 敲定生意的關鍵

Question
How do I request time off?　　　　　　該怎麼請假？

Answer

"Would it be possible to take Wednesday off next week? I have a doctor's appointment." The phrase "Would it be possible to V?" is considered to be a very polite way to ask for permission.

你可以說「Would it be possible to take Wednesday off next week? I have a doctor's appointment.」（請問下週三可以請假嗎？我得去看醫生。）其中「Would it be possible to V?」這種句型，是請求對方許可時相當有禮貌的一種表達方式。

請求安排出差事務
Requesting Travel Arrangements

當出差準備已徹底分工化且出差頻率很高時，就可將基本文章結構存成範本，這樣每當要出差時只需改寫日期、時間與目的地便可送出，非常方便。

Subject: L.A. trip

Latisha,

[1]I have to make a business trip next week to Los Angeles. [2]Please set up the reservations. [3]I want to fly JAL business class (leaving on Tuesday, returning on Saturday) and to stay at the Hilton (check in Tuesday, check out Saturday morning) in L.A. [4]Also, please arrange for a car and driver to pick me up at LAX to take me to the hotel, and since we will have a working dinner in L.A., please reserve a table for four at Largo for 7 p.m. on Wednesday night. [5]Any problems, give me a call.

Seiya

翻譯

主旨：到 L.A. 出差

Latisha：

我下週必須到洛杉磯出差，請幫我預約以下事項。我想搭 JAL 商務艙（週二去、週六回），並住 L.A. 的希爾頓飯店（週二入住、週六早上退房）。另外，請安排座車與司機到 LAX（洛杉磯國際機場）來接我到飯店，而由於我們在 L.A. 會有頓商務晚餐，故請幫忙在 Largo 餐廳訂位，週三晚上 7 點，共 4 人。如果有任何問題，請撥電話給我。

聖也

 重要片語 & 句型

- **make a business trip next week to ...** 下週到～出差
- **Please set up the reservations.** 請幫我預約。
- **LAX** 洛杉磯國際機場的別名（= Los Angeles International Airport）
- **Please arrange for ...** 請安排～。
- **working dinner** 商務晚餐（邊吃晚飯邊談生意）
- **Any problems, give me a call.** 如果有任何問題，請撥電話給我。

佳句便利貼

■ I have to go away on business to Zimbabwe.
我必須到辛巴威出差。

■ Please take care of the hotel in Miami for me.
請幫我預約邁阿密的旅館。

■ Can you do my rental car reservations for me?
能否請您幫我預約租車？

■ Please hire a limousine to pick me up at the airport.
請租一台小巴到機場接我。

■ A driver will meet you in the arrival's area for your transfer to the speaking venue.
有位司機會在入境大廳與您會合，並送您至演講地點。

FAQ 73　敲定生意的關鍵

Question

How can I request clarification of someone's meaning?

如何請對方針對其所說的話，做進一步說明？

Answer

"Did you mean ...?" is one of the easiest ways. For example, "When you said Tuesday, did you mean Tuesday this week or Tuesday next week?"

「Did you mean... ?」（你是指～嗎？）是最簡單的問法之一。例如，「When you said Tuesday, did you mean Tuesday this week or Tuesday next week?」（你說的週二，是指本週二還是下週二？）

74 出差報告
Reporting a Business Trip

出差後的報告也可做成專用範本，這樣只需改寫必要事項就能送出。如果有規定報告裡一定要寫上某些項目，或該企業有特定的文書格式時，就必須特別注意。

Subject: Last week's Frankfurt trip

T.K.,

[1]I'm writing to let you know the results of last week's business trip to our offices in Frankfurt. [2]I was provided with reports from their accounting, R&D team, and their management board and I had a look at the new product designs for the upcoming year. [3]I've just dropped off the performance reports and other relevant information from the trip to your secretary for your consideration. [4]If you need any other information about the trip, please let me know.

Keiko

翻譯

主旨：上週至法蘭克福的出差

T. K.：
在此寫信向您報告我上週到法蘭克福分公司出差的結果。他們的會計、研發團隊和管理委員會都提供了報告書給我，我也看了下一年度的新產品設計。我剛剛已經把出差所取得的績效報告和其他相關資訊交給您的秘書，供您評估。如果您還需要任何其他有關此次出差之資訊，請通知我一聲。

圭子

重要片語 & 句型

■ **I'm writing to let you know** ... 在此寫信向您報告～。
■ **be provided with** ... 提供了～給我
■ **have a look at** ... 看～（= look at）
■ **drop off** ... 將～交給

- **for** *one's* **consideration** 供～評估
- **If you need any other information about the trip, please let me know.**
  如果您還需要任何其他資訊，請通知我一聲。

佳句便利貼

- Several new ideas came up in the meeting that you may want to think about.
  此次會議中有幾個新點子，您或許會想考慮看看。
- I was offered an exclusive opportunity to invest in the new start-up.
  我獲得了投資那家新公司的獨家機會。
- I was given the opportunity to object to the new proposal.
  我有了反對那項新提案的機會。
- Please send me the address and any other information that might be useful.
  請將地址與任何其他可能用得到的資訊都寄給我。

FAQ 74　敲定生意的關鍵

Question
What does FYI mean?　　　　　　　　　FYI 是什麼意思？

Answer

FYI means "For Your Information." It means you may or may not want to know this but I'm going to tell you in case you are interested. It is commonly used when news stories and other general information memos are passed from employee to employee. "For your reference" is another expression to mean the same.

FYI 指「For Your Information」（供您參考）。這表示雖然此資訊你可能想也可能不想知道，但是為了以防萬一你有興趣，我還是通知你一聲。此縮寫經常用於員工之間相互轉寄的新聞或其他一般性備忘資訊中。而「For your reference」是另一種代表相同意義的講法。

74

出差費用請款
Requesting Reimbursement of Expenses

出差費用的計算很花腦力，公帳與私帳該如何劃分有時會讓人傷透腦筋，偶爾你甚至必須費力證明某些支出確實與出差工作相關。

Subject: India expense report

Sanja,

[1]Here are the receipts and my expense report from my trip to India. [2]The two restaurant receipts are both from business dinners I had with TYA Outsourcing and P.K. Industries. [3]Please understand that in some cases, such as public transportation, receipts simply were not possible. [4]Once the paperwork is done, please inform me of the reimbursement date. [5]Thank you.

Kosuke Maruyama

翻譯

主旨：印度出差費用報告

Sanja：
這些是我去印度出差時的收據和支出報告。其中的兩張餐廳收據，是與 TYA 供應商及 P.K. 工業公司共進商務晚餐的費用。在某些情況下，例如搭乘大眾交通工具時，是無法取得收據的，這點尚請見諒。書面作業一旦完成，請通知我退款日期。謝謝您。

丸山幸助

 重要片語 & 句型

■ **expense report** 費用報告
■ **in some cases such as ...** 在某些情況下，例如～
■ **Please understand that ...** ～，這點尚請見諒。
■ **Once the paperwork is done, please inform me of the reimbursement date.**
　書面作業一旦完成，請通知我退款日期。

佳句便利貼

■ What are your expenses for transportation?
你的交通費是多少？

■ This is a list of the costs of this year's utilities.
這是本年度的水電瓦斯費清單。

■ Please recognize that reducing labor costs is part of business.
請體認降低人力成本也是營運管理的一環。

■ Could you pay me back the money you owe me?
能否請您把欠我的錢還給我？

 FAQ 75 敲定生意的關鍵

Question
How can I dispute a reimbursement? 該如何爭取退款？

Answer

Businesses often don't want to pay the real cost of a business trip, so arguing is important to get reimbursed properly. "I would like to dispute the amount you reimbursed me for my meal expenses. These two dinners were business meetings with clients. I believe the company should be responsible for the cost." Using these phrases will at least let the company know that you are not pleased with their policies.

企業往往不願意支付出差的實際費用，因此如果想拿回應有的退款，據理力爭相當重要。使用「I would like to dispute the amount you reimbursed me for my meal expense. These two dinners were business meetings with clients. I believe the company should be responsible for the cost.」（我對餐費的退款金額有異議。這兩頓晚餐都屬於和客戶的商務會談，我認為公司應該要負擔這項費用。）這類措辭至少能讓公司了解你對他們的處理方式並不滿意。

9

個人問候・祝賀
Personal Greetings / Congratulations

76 祝賀聖誕節
Christmas Greetings

傳統上，聖誕節時人們都會郵寄卡片相互祝賀佳節愉快。但是近年來愈來愈多人改用 email 祝賀，有時還會附上電子賀卡。一般來說，在沒有寄送賀年卡習慣的歐美等地，多半會將聖誕節和新年一起合併祝賀。

Subject: Happy Holidays

To the Harris family,

[1]Happy Holidays! [2]Wishing you and your family a very merry Christmas and a happy new year!

Osamu Takai

翻譯

主旨：佳節愉快

給 Harris 家族：
佳節愉快！祝福您和全家聖誕快樂、新年快樂！

高井修

 重要片語 & 句型

■ **Happy Holidays!** 佳節愉快！
■ **Wishing you (and your family) a very merry Christmas and a happy new year!**
　　祝福您（和您全家）聖誕快樂、新年快樂！

76

佳句便利貼

- To the Allens,
  給 Allen 全家：

- To Tom, Sylvia and the kids,
  給 Tom、Sylvia 以及孩子們：

- Season's Greetings!
  聖誕快樂！（用於賀卡上）

- Best wishes during this holiday season!
  在這佳節期間為您獻上最深祝福！

- Here's hoping for the success of each of you.
  在此敬祝各位馬到成功。

- Have a wonderful 20XX!
  祝您有個美好的 20XX 年！

- I hope this year brings you everything you desire.
  祝您今年心想事成。

 **FAQ 76** 敲定生意的關鍵

Question

Why "Wishing you a happy new year!" and not "Wishing you Happy New Year!"?

為什麼說「Wishing you a happy new year!」而不是「Wishing you Happy New Year!」？

Answer

In "Wishing you a happy new year!" the meaning is not the same as the expression "Wishing you Happy New Year!". The first means a wish for the entire year, not just the greeting said around New Year's Day. "Wishing you Happy New Year!" is awkward and not natural English. If the meaning is the holiday expression people say, then it is easier just to say "Happy New Year!"

「Wishing you a happy new year!」所表達的意思和「Wishing you Happy New Year!」並不相同。前者是祝福對方一整年，而不只是針對新年假期表達祝賀。說「Wishing you Happy New Year!」有些奇怪，就英語表達而言不太自然。如果是想針對假期表達慶賀之意，直接說「Happy New Year!」就可以了。

77 祝賀新年
New Year Greetings

用賀年卡祝賀新年的方式在歐美並不普遍,但倒是有利用 email 祝賀新年的做法。如果你的祝福對象身處中、韓、日等十二生肖文化深植人心的國家,那麼也可提及生肖話題。

Subject: Happy New Year!

Trisha,

[1]Happy New Year! [2]20XX was a great year for the both of us and I hope that 20YY brings you even more happiness and success!

Ichiro

翻譯

主旨:新年快樂!

Trisha:

新年快樂! 20XX 年對我們來說都是很棒的一年,我希望 20YY 年能為您帶來更多快樂與成就!

一朗

 重要片語 & 句型

■ **Happy New Year!** 新年快樂!

■ **I hope that** ... **brings you even more happiness and success.**
  我希望~能為您帶來更多快樂與成就!

77

佳句便利貼

- **Happy 20XX!**
  20XX 年新年快樂！

- **New Year's wishes to your family!**
  為您全家獻上新年祝福！

- **I hope this year welcomes you with much joy.**
  祝您新的一年充滿歡樂。

- **20XX will bring you much prosperity.**
  祝您 20XX 年事業興旺。

- **I hope you will get everything your heart desires.**
  祝您心想事成。

- **You and I both will find work this year.**
  我們今年都能求職成功。

- **The two of us are planning to open a business together.**
  我們兩個計畫一同開創新事業。

FAQ 77　敲定生意的關鍵

Question
How can I refuse an invitation?　　　　該如何拒絕別人的邀請？

Answer

There are a few different ways. For example, "I'm afraid I'm busy." is OK. Or if it's something you don't enjoy doing like clubbing, you might say, "Thanks. But I'm not really a club person."

有幾種不同的說法。例如你可以說：「I'm afraid I'm busy.」（我恐怕有點忙。）而如果是你不喜歡做的活動，像是去夜店之類的邀約，那就可以說「Thanks. But I'm not really a club person.」（謝謝，我真的不愛泡夜店。）

祝賀情人節
Valentine's Day Greetings

在台灣，情人節時並沒有像日本一樣，由女生送巧克力給男生的習慣。而在歐美則通常是由男性送巧克力等禮物給女性，還有很多男人會在報紙上刊登送給妻子的愛的留言呢！

Subject: Happy Valentine's Day!

Beth,

[1]Happy Valentine's Day, Sweetheart! [2]I'm thankful for every day with you! [3]I'm so happy we met! [4]You are the only one for me!

I love you!

Nobu

翻譯

主旨：情人節快樂！

Beth：
親愛的，情人節快樂！我每天都心存感激，因為能和妳在一起。我好高興我們能相遇！妳是我的唯一！

我愛你！
信

重要片語 & 句型

■ **Happy Valentine's Day!** 情人節快樂！
■ **sweetheart** 甜心；親愛的
■ **I'm thankful for every day with you!** 我每天都心存感激，因為能和你在一起！
■ **I'm so happy we met!** 我好高興我們能相遇！
■ **You are the only one for me!** 你是我的唯一！

- Happy Valentine's Day, my love!
  親愛的,情人節快樂!

- Happy Valentine's Day to the love of my life!
  情人節快樂,我的一生摯愛!

- Every day with you is special!
  和你在一起的每一天都很特別!

- Each day with you makes me happy!
  和你在一起的每一天都讓我無比幸福!

- You are my everything!
  你是我的一切!

- With all my love,
  獻上我滿滿的愛。(用於信中結尾)

- I love you with all my heart.
  我全心全意地愛你。

FAQ 78　敲定生意的關鍵

Question
How do I suggest an alternative to what has been suggested already?

針對已決定事項,該如何提出一個不同的替代方案?

Answer
For example, if someone invites you out for a steak dinner but you are a vegetarian, you might say something like, "A restaurant sounds great, but how about somewhere with a salad bar?"

舉例來說,如果有人邀請你去吃牛排晚餐,而你吃素,那就可以說「A restaurant sounds great, but how about somewhere with a salad bar?」(去餐廳很棒,但是去有沙拉吧的地方如何?)

祝賀生日的 email 非常重要,且依對方年齡不同,該用的賀詞也不一樣。

Subject: Happy Birthday!

Tori,

[1]Happy Birthday! [2]With age comes wisdom! [3]I'm sure you are aging gracefully. [4]I hope this year brings you everything you wish for and that there are many more birthdays to come!

Tsuyoshi

翻譯

主旨:生日快樂!

Tori:
生日快樂!祝妳添歲長智慧!我確信隨著年齡增長,妳變得愈來愈優雅。祝妳今年事事順心,並且福如東海、壽比南山!

剛

重要片語 & 句型

■ **Happy birthday!** 生日快樂!
■ **With age comes wisdom!** 添歲長智慧!
■ **I'm sure you are aging gracefully.** 我確信隨著年齡增長,你變得愈來愈優雅。
■ **I hope this year brings you everything you wish for.** 祝你今年事事順心。
■ **I hope that there are many more birthdays to come.** 祝你福如東海、壽比南山!

佳句便利貼

■ I hope you find everything you're looking for.
祝您萬事如意。

■ Your best years are ahead of you!
您意氣風發的年代即將來臨！

■ The best is yet to come!
最棒的才剛要開始！

■ You are aging like a fine wine.
您就像瓶好酒，愈陳愈香。

■ Age brings experience.
歲月豐富了經歷。

■ You've aged beautifully.
您成熟中更見美麗。

■ You still look as young as ever!
您依舊青春如昔！

 FAQ 79 敲定生意的關鍵

Question

Is it OK to ask someone how old they are on their birthday?

在他人生日的時候，是否可詢問他們的年齡？

Answer

It depends on the person. Women often don't like to talk about their age, and more and more some men feel the same way. Age is considered less important in many western cultures than in Asia.

這因人而異。女性通常不喜歡談論年齡，而且有愈來愈多的男性也是如此。和亞洲人不同的是，在許多西方文化中並不很重視年齡大小。

畢業祝賀
Celebrating Graduation

恭賀畢業的 email 可以寄給高中、大學畢業的年輕人，以及從研究所畢業取得了碩士或博士學位的人。當然，讀在職班而取得 MBA（Master of Business Administration）學位的人也可列為祝賀對象。

Subject: Congratulations on your graduation!

Briana,

[1]Congratulations on your graduation from college! [2]The many years you have spent studying will soon pay off. [3]I wish you much success in your future endeavors!

Sachiyo Mima

翻譯

主旨：恭喜您順利畢業！

Briana：
恭喜妳大學畢業了！多年苦讀即將有所回報。祝妳未來的努力都能成功！

三間幸代

 重要片語 & 句型

- **Congratulations on ...!** 恭喜你～！
- **graduation from college** 大學畢業
- **The many years you have spent ... will soon pay off.** 你多年的～即將有所回報。
- **I wish you much success in your future endeavors!** 祝你未來的努力都能成功！

佳句便利貼

- You graduated! Congratulations!
  你畢業了！恭喜！

- Congratulations on your getting an MBA!
  恭喜您取得 MBA 學位！

- Your hard work has finally come to an end.
  您辛苦的努力總算告一段落。

- Many years of effort have led to this moment.
  經過多年努力才有這一刻。

- You will soon be rewarded for all that you have done in life.
  您這一生中的所有付出即將得到回報。

- You certainly have a bright future ahead of you!
  你的前途肯定一片光明！

- Good luck in your future!
  祝你未來一路好運！

 FAQ 80 敲定生意的關鍵

Question

What do I say if I forgot about someone I was supposed to meet today?

我今天本應與某人會面，但是卻忘了這件事。我該說些什麼？

Answer

Apologize quickly and explain what happened starting with "I'm really sorry. I completely forgot." In this case an email message is usually not considered appropriate and it is preferable to call them to apologize directly to them.

請以「I'm really sorry. I completely forgot.」（真的很抱歉，我完全忘了。）這句起頭，趕快道歉並解釋原因。在這種情況下並不適合用 email，打電話直接道歉比較妥當。

無論是公司內還是公司外的工作夥伴，只要有人升官，都應寄送祝賀的 email。信中不應只寫一句恭喜的話，順便讚美一下對方的努力會更好。

Subject: Congratulations on your promotion!

Taylor,

[1]Congratulations on your big promotion to management! [2]It's been a long time coming and you are very deserving. [3]You must be excited. [4]Great job! [5]I'm sure you will make an excellent manager.

Koji Noda

翻譯

主旨：恭喜你升官了！

Taylor：
恭喜你擢升管理階層！真是期待已久，而你也當之無愧。你一定很高興吧。做得好！我相信你一定會成為一位優秀的經理。

野田浩二

 重要片語 & 句型

■ **Congratulations on your promotion.** 恭喜你擢升。
■ **management** 管理階層
■ **It's been a long time coming.** 真是期待已久。
■ **You are very deserving.** 這是你應得的；你當之無愧。（= You deserve it.）
■ **You must be excited.** 你一定很高興吧。
■ **Great job!** 做得好！（= Good job! / Good work! / Great work! / Way to go!）
■ **I'm sure you will make ...** 我相信你一定會成為～。

佳句便利貼

■ Congrats on getting promoted!
恭喜你獲得升遷！

■ Congratulations on your rise to the top of the company.
恭賀您榮升公司最高階層。

■ It has taken a long time, but you made it to Tokyo.
雖然花費了不少時間，但是您終於將事業拓展至東京了。

■ It was worth the wait to finally reach your goal of managing a professional soccer team.
您終於達成目標成為了職業足球隊總教練，一切的等待都是值得的。

■ You deserve it!
這是你應得的！

■ You are worthy of every success that comes your way.
您的成功全都實至名歸。

■ I have no doubt you will be a great boss.
我確信你一定會成為一個傑出的老闆。

 FAQ 81 敲定生意的關鍵

Question

How can I ask someone to reconsider their idea?

該如何要求對方重新考慮他們的想法？

Answer

You can say, "I think you should reconsider your plan/idea." Asking someone to reconsider their idea may sometimes cause an argument. Try to be as kind and friendly as possible when using this phrase.

你可以說「I think you should reconsider your plan / idea.」（我覺得你該重新考慮一下你的計畫／想法。）要求對方重新考量有時會引發爭執，因此在使用這類措辭時，應盡量表現出親切友善的態度。

退休祝賀
Celebrating Someone's Retirement

有些國家（如日本）會將達到一定年齡便退休視為理所當然，但是也有像美國這類基本上並無屆齡退休習慣的國家。對非自願退休的人表達祝賀是非常不恰當的，這點請特別注意。

Subject: Enjoy your retirement!

Barney,

[1]Congratulations on your retirement! [2]I was sad to hear that you will be leaving your company. [3]I have enjoyed working with you over these past many years. [4]I hope you will have a wonderful retirement and that you can now do all those things you've dreamed of doing. [5]Best wishes and continued happiness in this new chapter of your life.

Yoshihisa Sakurai

翻譯

主旨：好好享受您的退休生活！

Barney：
恭喜您退休了！聽到您即將離開貴公司，我很感傷。過去多年來與您一起工作非常愉快。希望您能有美好的退休生活，並能做所有您夢想要做的事。在此獻上最大的祝福，願您在這人生新的一章中繼續過著幸福快樂的日子。

櫻井義久

重要片語 & 句型

■ **Congratulations on your retirement!** 恭喜您退休了！
■ **I was sad to hear that you will be leaving your company.**
　聽到您即將離開貴公司，我很感傷。
■ **I have enjoyed working with you.** 過去與您一起工作非常愉快。
■ **over these past many years** 過去多年來
■ **I hope you will have a wonderful retirement.** 希望您能有美好的退休生活。

■ **You can now do all those things you've dreamed of doing.**
您現在可以做所有您夢想要做的事。

■ **Best wishes and continued happiness in this new chapter of your life.**
在此獻上最大的祝福,願您在這人生新的一章中繼續過著幸福快樂的日子。

佳句便利貼

■ Retirement is just another phase in your life.
退休只是您人生的另一個階段。

■ As you start this new part of your life, please continue learning your trade.
隨著您開啟人生的新篇章,請繼續充實您的手藝。

■ I hope you can realize your dream of opening a chain of cafes.
希望您能實現開連鎖咖啡廳的夢想。

■ It was quite a shock to hear that you quit your job.
聽到您辭職的消息,令我相當震驚。

 **FAQ 82** 敲定生意的關鍵

Question

I got an email attached with a file but I cannot open it. I am a Mac user and the sender is a Windows user. How can I ask them not to attach anything to message sent to me or ask them to send the contents inside the email message?

我收到一封有附件的 email,但是打不開附件檔案。我用的是 Mac 電腦,而寄信人用的是 Windows 系統。我該怎麼告訴他們寄給我的信不要加上附件,或是要求他們將要傳送的內容直接貼在 email 內文中?

Answer

You could write, "I'm afraid I can't open your attachment. Please resend it inside the body of the email message." This should work.

建議你可以這樣寫:「I'm afraid I can't open your attachment. Please resend it inside the body of the email message.」(抱歉,我無法開啟您寄來的附檔。麻煩將該附件內容貼在 email 內文中,重寄一次。)這樣應該就行了。

對於剛搬到自家附近的人，可以用 email 表達歡迎之意，甚至邀請對方到家裡做客。如果無法取得對方的 email 地址，就改採口頭祝賀或遞送卡片的方式。

Subject: Welcome

Bill and Sue,

[1]Welcome to the neighborhood! [2]We live in the house next door (to your left). [3]We'd like to have you over for a drink this weekend if you are free. [4]This is a great neighborhood, with nice people. [5]We hope you'll enjoy living here.

Xavier and Mina Linney

翻譯

主旨：歡迎

Bill 與 Sue：
歡迎搬到本社區！我們就住在隔壁（你們左邊）。如果兩位有空，我們想邀請你們週末過來喝一杯。這是個很棒的社區，大家都很和善。希望你們在這兒住得愉快。

Xavier 與 Mina Linney

 重要片語 & 句型

■ **Welcome to the neighborhood!** 歡迎（搬）到本社區！
■ **We live in the house next door.** 我們就住在隔壁。
■ **We'd like to have you over for a drink.** 我們想邀請你們過來喝一杯。
■ **if you are free** 如果你們有空
■ **We hope you'll enjoy living here.** 希望你們在這兒住得愉快。

■ Welcome to the area!
歡迎來到本區！

■ Welcome to town!
歡迎來到鎮上！

■ The office you want is in the building next door.
你要找的辦公室在隔壁棟。

■ Why don't you stop by for a drink later?
您何不稍後過來喝一杯？

■ Care to join us for a drink at my place?
願不願意到我家來和我們喝一杯？

■ If you have some time, can we go to the flea market together?
如果你有空，要不要一起到跳蚤市場逛逛？

 **FAQ 83** 敲定生意的關鍵

Question
How do I say " 好久不見 " in English?　　　　「好久不見」的英文怎麼說？

Answer

"It's been a long time!" or "I haven't seen you for a long time!" We use these phrases in very much the same way as Chinese people use " 好久不見."

「It's been a long time!」或者是「I haven't seen you for a long time!」這些表達方式都和中文的「好久不見」差不多。

祝賀訂婚一事也能透過 email 完成。以下範例發生在發信者從第三者聽到消息，而非直接接獲當事人通知的狀況下。在這類信中除了詢問婚禮等資訊外，更需傳達出由衷的喜悅。

Subject: You're engaged!

Ron and Jenna,

[1]I was so happy when I got your announcement that you two got engaged! [2]Congratulations! [3]When is the wedding and where will it be held? [4]I wish the happy couple a long life together. [5]Best wishes to you both!

Junya

翻譯

主旨：你們訂婚了！

Ron 與 Jenna：

得知你們兩位訂婚的消息時，我真的非常高興！恭喜！婚禮將於何時、何地舉行？願兩位佳偶天成、永浴愛河。在此獻上最美好的祝福！

純也

 重要片語 & 句型

■ **get engaged** 訂婚
■ **When is the wedding and where will it be held?** 婚禮將於何時、何地舉行？
■ **I wish the happy couple a long life together.** 願兩位佳偶天成、永浴愛河。
■ **Best wishes to you both!** 在此為兩位獻上最美好的祝福！

佳句便利貼

■ I heard you two are tying the knot.
我聽說你們倆打算結婚。

■ Where are you going to have the ceremony?
你們的婚禮將在哪裡舉行？

■ I wish the future Mr. and Mrs. Smith a happy future together.
祝未來的 Smith 夫婦幸福美滿、永浴愛河。

■ I wish the soon-to-be-newlyweds all the happiness in the world.
恭祝即將結合的新婚夫妻們幸福無限。

 **FAQ 84** 敲定生意的關鍵

Question
How can I express surprise at someone's happy news they email me in English?

對於他人以英文 email 傳來的好消息，我該如何表達驚喜之意？

Answer
"Wow!" "That's great!" "That's good news!" "That's great to hear!" are a few of the ways you can express surprise for someone's happy news.

「Wow!」、「That's great!」、「That's good news!」和「That's great to hear!」等幾句話，都可用來表達對好消息的驚喜之意。

結婚祝賀與訂婚祝賀一樣，務必表達出由衷的祝福。但比言語更重要的是，這種 email 愈早送出愈好，因為第一封寄到的祝福，最令人感到貼心。

Subject: To the happy couple!

Tran and Stacy,

[1]My heartfelt congratulations on your getting married! [2]You two were meant for each other! [3]I wish you much happiness and joy together as you start your lives as husband and life.

Eri

翻譯

主旨：給幸福佳偶！

Tran 與 Stacy：
誠摯地恭喜兩位新婚快樂！你們倆真是天生一對！願你們的婚姻生活充滿幸福與喜悅。

惠理

重要片語 & 句型

■ **My heartfelt congratulations on your getting married!** 誠摯地恭喜兩位新婚快樂！
■ **You two were meant for each other!** 你們倆真是天生一對！
■ **I wish you much happiness and joy together.** 願你們在一起幸福滿滿、喜悅滿滿。

佳句便利貼

■ Most sincere congratulations are in order.
獻上最誠摯的祝福。

■ Congratulations from the bottom of my heart.
從我心底獻上最深切的祝賀。

■ I'm so excited about your recent marriage to Trinity.
聽到你最近和 Trinity 結婚了，我非常高興。

■ You two are perfect together.
你們倆是天作之合。

■ Enjoy your new life as a married couple.
好好享受你們的新婚生活。

 FAQ 85 敲定生意的關鍵

Question

If I'm invited to someone's home for dinner, what is the polite etiquette for accepting by email?

獲邀前往某人家中吃晚餐時，該如何以 email 有禮貌地表示接受邀請？

Answer

You can say "It would be my pleasure to come." This is the common acceptance phrase. You should always ask "Can I bring something?." This question is polite etiquette and refers to things like food or beer or a bottle of wine or something to add to the dinner.

你可以說：「It would be my pleasure to come.」（我很樂意前往。）這是接受邀約時最常見的說法。記得一定要問：「Can I bring something?」（我能帶點什麼過去嗎？）此句話非常客氣且符合禮儀，意思是你想帶其他食物或啤酒、紅酒等使晚餐更豐盛些。

祝賀生產
Celebrating the Birth of a Child

祝賀生產的 email 也是愈早送出愈好。愈晚寄出，效果愈差，給人的印象也不好。而信中除了祝福小孩健康成長外，也應慰勞一下當事人育兒的辛勞。

Subject: Your new baby!

Tim and Jessica,

[1]Congratulations on the birth of your new baby girl! [2]May your new daughter grow up healthy and happy. [3]I'm sure you'll both be the best parents a baby could ask for. [4]Enjoy parenthood! [5]They grow up quick, so cherish every moment!

Rika

翻譯

主旨：你們的新寶寶！

Tim 與 Jessica：

恭喜你們的女寶寶誕生！願你們的女兒健康快樂地成長。我相信你們一定會成為寶寶能擁有的最佳父母。請好好享受育兒時光！孩子的成長非常迅速，所以請務必珍惜每一刻！

理香

 重要片語 & 句型

■ **Congratulations on the birth of your new baby girl / boy!** 恭喜你們的女／男寶寶誕生！
■ **May your new daughter / son grow up healthy and happy.**
　願你們的女兒／兒子健康快樂地成長。
■ **I'm sure you'll both be the best parents a baby could ask for.**
　我相信你們一定會成為寶寶能擁有的最佳父母。
■ **Enjoy parenthood!** 請好好享受育兒時光！
■ **A child grows up quickly, so cherish every moment!**
　孩子的成長非常迅速，所以請務必珍惜每一刻！

佳句便利貼

■ May you find happiness.
　願您找到幸福。

■ May you be happy in your future.
　願您未來幸福快樂。

■ Enjoy every minute of your parenthood!
　請享受育兒時的每一刻！

■ Yoshiko is the woman of my dreams.
　芳子是我夢想中的女性。

 FAQ 86 敲定生意的關鍵

Question
How do I say "沒辦法" or "無能為力" in English?

「沒辦法」或「無能為力」的英文該怎麼說？

Answer

These phrases both roughly mean "It can't be helped." or "There's nothing we can do about it." The big difference, however, is that English speakers only use these expressions for situations where truly nothing can be done. Taiwanese tend to use "沒辦法" and "無能為力" for situations where something can be done. Be careful when you use these in English or people may think you are lazy.

這兩種說法的意思差不多等於「It can't be helped.」或者是「There's nothing we can do about it.」，而主要的差異在於，以英語為母語的人只在真的無計可施的情況下才會用這些說法，但台灣人在「尚且有計可施」的情況下仍習慣把「沒辦法」或「無能為力」這兩句話掛在嘴邊。因此，在使用這類英文句子時務必謹慎，否則別人會覺得你很懶散。

87 敬祝早日康復
Get-well Messages

一般而言，臥病在床的人對於那些曾表達關心的人永遠都會心懷感謝。而相反地，即使平常自己和人家交情不錯，但對方生病時卻連封慰問的 email 都不寄，這樣恐怕會給人相當負面的觀感。

Subject: Get well soon!

Tess,

[1]I was so sorry to hear that you are in the hospital. [2]I hope you'll be on the mend soon and will be out of the hospital quickly.

All the best,

Zaizen

翻譯

主旨：祝您早日康復！

Tess：
很遺憾聽說您住院了。希望您很快就可以康復出院。

祝福您。
Zaizen

 重要片語 & 句型

■ **I was so sorry to hear that you are in the hospital.** 很遺憾聽說您住院了。
■ **I hope you'll be on the mend soon.** 希望您很快可以康復。
■ **I hope you'll be out of the hospital quickly.** 希望您很快就能出院。
■ **All the best,** 祝福您。

佳句便利貼

■ I hope you'll make a quick recovery from your illness.
希望您的病能迅速痊癒。

■ Best wishes!
祝福您！

■ Get well soon!
早日康復！

■ James was hospitalized with pneumonia last night.
James 昨晚因肺炎入院。

■ Susumu was admitted to the hospital last night after complaining of chest pain.
Susumu 昨晚在抱怨胸痛後，就住進了醫院。

 FAQ 87 敲定生意的關鍵

Question

How do I say " 請多保重 " in English?　　　　　「請多保重」的英文該怎麼說？

Answer

"Take care." or "Get better soon." are two of the common ways to say " 請 多 保 重 " in English. In addition, "Take good care of yourself." is used when you express more concern for someone.

「Take Care.」或「Get better soon.」是英文中兩種常見可用來表達「請多保重」之意的說法。另外，「Take good care of yourself.」（好好照顧你自己。）則可用來表達更進一步的關懷。

喪事弔唁極為重要，用字遣詞時需特別小心謹慎。而且不只針對人，有時連寵物過世也需寄 email 表達哀悼之意。接下來的範例，就是針對朋友愛貓過世而寫的弔唁信。

Subject: My condolences

Tala,

[1]My deepest sympathies about the loss of your beloved cat, Sasha. [2]What a wonderful companion she was for all these years.

Daigo

翻譯

主旨：我的哀悼之意

Tala：

對於妳痛失愛貓 Sasha，我在此獻上最深切的哀悼之意。這些年來，她真是妳最棒的生活伴侶。

大悟

 重要片語 & 句型

■ **My deepest sympathies about the loss of** ...
　對於你痛失～，我在此獻上最深切的哀悼之意。

■ **wonderful companion** 最棒的生活伴侶

■ **for all these years** 這些年來

佳句便利貼

- My condolences on the loss of your friend.
  在此為您痛失好友一事致上哀悼之意。

- I share your grief about the passing of Mr. Oldham.
  對於 Oldham 先生過世，我與您同感悲傷。

- My dog was a great partner in my life.
  我的狗兒曾是我生活中的絕佳夥伴。

FAQ 88　敲定生意的關鍵

Question

Someone contacted me on a SNS but I can't remember how I know them. How can I check in English?

有人在社群網站上跟我聯繫，但是我記不起來是怎麼認識這些人的。此時該如何用英文向對方確認？

Answer

You might ask "Refresh my memory. How do we know each other?" It's sometimes difficult to tell who knows you and who is contacting you for the first time. Alternatively, if you are worried about who the person contacting you is, it may be better to ignore the message.

你可以問「Refresh my memory. How do we know each other?」（請提醒我一下，我們到底是怎麼認識的？）。要分辨誰認識你，誰又是第一次與你接觸，有時並不容易。如果對與你接觸的人有所疑慮，那麼忽略對方來信的做法是最保險的。

寄送賀禮時，多半會一併附上賀卡，不過你也可只寄送禮物，再於對方收到前以 email 表達慶賀之意。

Subject: A little something for you

Sylvie,

[1]I just sent a package over to you in celebration of your birthday. [2]It's a little something I picked up on my recent trip to Tahiti. [3]I hope you like it. [4]It's something you've always wanted.

Kiyoto

翻譯

主旨：給你的一點小心意

Sylvie：
為了祝賀妳的生日，我剛剛寄了個包裹過去給妳。裡面是我最近去大溪地旅遊時挑選的小東西。希望妳會喜歡。那是妳一直都很想要的東西。

清人

 重要片語 & 句型

■ **I just sent a package over to you in celebration of** ...
為了祝賀～，我剛剛寄了個包裹過去給你。

■ **It's a little something I picked up on my recent trip to** ...
那是我最近去～旅遊時挑選的小東西。

■ **I hope you like it.** 希望你會喜歡。

■ **It's something you've always wanted.** 那是你一直都很想要的東西。

佳句便利貼

- Let's open a bottle of champagne to celebrate your promotion.
  咱們開瓶香檳來慶祝你高升。

- Here's a watch to mark this special occasion.
  謹以此手錶來慶祝這個特別的時刻。

- This is a small token of my appreciation.
  這個小東西代表了我的謝意。

- I hope these flowers are to your liking.
  希望你喜歡這些花。

- I hope this vase suits your tastes.
  希望這個花瓶你會喜歡。

- Here is the necklace you've been longing for.
  這是你渴望已久的項鍊。

- I found the book you've been wanting for a long time.
  我找到了你一直想要的那本書。

 FAQ 89 敲定生意的關鍵

Question

I've lost something and I would like to send a message to my friend to see if he/she has it. What can I say?

我弄丟了某個東西,想寫封信向朋友確認是否在他/她手上。我該怎麼表達?

Answer

"Have you by chance seen my key anywhere?" or "You didn't happen to find my wallet anywhere, did you?" Both of these are indirect ways of asking and are seen as polite.

「Have you by chance seen my key anywhere?」(你會不會碰巧看到了我的鑰匙?)或者是「You didn't happen to find my wallet anywhere, did you?」(你不會剛好找到了我的皮夾吧?)這兩種問法都比較委婉,感覺也較有禮貌。

針對他人送來的禮物要表達謝意時，如果彼此原本就是經常以 email 聯絡的朋友，那麼用 email 致謝也無所謂。而信中如果能寫上你如何運用了該份禮物，就肯定能讓送禮者更加高興。

Subject: The beautiful flowers

Nancy,

[1]Thank you so much for the beautiful flowers you sent. [2]I have them in a vase on my mantle and they really brighten my day. [3]You are so thoughtful and considerate.

翻譯

主旨：美麗的花

Nancy：
非常感謝妳送的美麗花朵。我把它們插進了花瓶裡放在壁爐台上，這些花真的讓我一整天都心情開朗。妳真是體貼又周到。

 重要片語 & 句型

■ **... really brighten my day.** ～真的讓我一整天都心情開朗。
■ **You are so thoughtful and considerate.** 你真是體貼又周到。

佳句便利貼

■ Thanks for the beautiful handmade card you sent!
謝謝您寄來的美麗手工卡片！

■ I appreciate the delicious apple pie you sent over.
很感謝您送過來的美味蘋果派。

■ Roses always make my day.
玫瑰花總是能讓我一整天都有好心情。

■ A phone call from you always makes me happy.
接到你的電話總是讓我很開心。

■ You are always thinking of me!
你總是為我設想！

■ Thank you for considering my feelings.
謝謝你顧慮到我的感覺。

■ Thank you for keeping my father in your thoughts while he is in the hospital.
謝謝你在我父親住院時仍掛念著他。

 FAQ 90 敲定生意的關鍵

Question

Sometimes I get email messages with Xs and Os as in "XX" or "XOXO." What do these mean?

有時我會收到寫了幾個 X 和 O 字母的信，例如「XX」或「XOXO」等。這些字母代表了什麼意思？

Answer

Xs represent kisses and Os represent hugs. We use these as closings to love interests and very close friends in English. XX means a couple of kisses and XOXO means kisses and hugs.

X 代表親一下，而 O 代表擁抱。在英文郵件中，結尾時我們會對戀愛對象或親密好友使用這些字母。XX 代表親兩下，XOXO 就是又親又抱。

10

私人事務通知
Personal Announcements

91 電話號碼・地址變更通知
Announcing a New Phone Number or New Address

一旦搬家，就需馬上告知同事及客戶新地址和新的聯絡方式。此時可直接將同一封 email 寄給多名收件者。英文地址的寫法和中文的寫法順序相反，這點請特別留意。

Subject: Change of address

Dear friends,

[1]This is to let everyone know that I have moved from Madison to Carbondale and to provide you with my new details. [2]I've just moved into my new apartment. [3]My new address is 408 S. Ash, Carbondale, Illinois 62901. [4]My new telephone number is 618-453-7528. [5]Have a great day!

Momoko

翻譯

主旨：住址變更

親愛的朋友們：
在此通知各位親朋好友我已經從 Madison 搬到了 Carbondale，同時我也要提供新的詳細聯絡資訊。我剛搬進新公寓裡，新住址是 408 S. Ash, Carbondale, Illinois 62901，新電話號碼為 618-453-7528。祝你有個美好的一天！

桃子

 重要片語 & 句型

■ **This is to let everyone know that I have moved from *A* to *B* and to provide you with my new details.**
在此通知各位親朋好友我已經從 A 搬到了 B，同時我也要提供新的詳細聯絡資訊。

■ **Have a great day!** 祝你有個美好的一天！

佳句便利貼

■ Here's my address and phone number in case you need it.
以防萬一你需要，這是我的地址與電話號碼。

■ I'm renting an apartment on the Dole Street.
我在 Dole 街上租了一間公寓。

■ I'm living in temporary housing until I can find something permanent.
在找到可永久居住的地方之前，我現在暫時租房子住。

■ Give me your contact information so I can let you know my new address.
請給我您的聯絡資訊，我好通知您的新地址。

 FAQ 91 敲定生意的關鍵

Question
What does "lol" mean? 「lol」是什麼意思？

Answer
It means "laugh out loud." In email messages between friends we use "lol" after a funny sentence.

「lol」代表「laugh out loud」（大笑出聲）之意。在寫給朋友的信中，大家經常會在有趣的句子後加上「lol」。

搬家通知
Announcing a Move

搬家通知也可在搬家前就先行送出。於搬家時把不需要的東西以低價售出稱為 rummage sale（清倉拍賣），這做法在美國相當普遍，可省下處理廢棄物的麻煩與費用。在下面這個例子中就提到了 rummage sale。

Subject: We're moving!

---

[1]We've just signed the papers on our new house in nearby Point Place. [2]We should be at our old address until the end of this month and will be having a rummage sale on the 26th and 27th. [3]The movers are coming on the 29th. [4]Our new address will be 1211 Armstrong Avenue. [5]We'll let you all know when the housewarming party will be.

Red and Kitty Forman

翻譯

主旨：我們要搬家了！

我們剛簽了就在附近 Point Place 的新房子合約。至本月底為止，我們應該都還在舊地址處，而我們將於 26 與 27 日舉辦舊物清倉拍賣。搬家公司 29 日會來。新住址是 1211 Armstrong Avenue。我們會再通知大家喬遷派對將於何時舉辦。

Red and Kitty Forman

 重要片語 & 句型

■ **We're moving!** 我們要搬家了！
■ **sign the papers on** ... 簽～合約
■ **We should be at our old address until** ... 至～為止，我們應該都還在舊地址處。
■ **rummage sale** 舊物清倉拍賣
■ **mover** 搬家公司
■ **housewarming party** 喬遷派對

佳句便利貼

■ We just closed on our new condo.
我們剛完成了新大樓住宅的購屋手續。

■ Our house just went into escrow.
我們的房子剛剛交給了交易履約代管。

■ Pete and Martha are having a moving sale.
Pete 和 Martha 正在舉辦搬家拍賣。

■ Which moving company did you use when you changed houses?
你搬家時用了哪一家搬家公司？

■ When are we christening your new house?
我們何時可到你的新家參觀？

■ When's the open house for your new home?
你的新家何時開放參觀？

 FAQ 92 敲定生意的關鍵

Question
I need to give someone directions to my office in English. What are some expressions I can use?

我必須以英文指引某人到我的辦公室。我可以用哪些表達方式？

Answer
You can say "Turn left," or "Turn right." "Go straight." "It's the building next to the police station." Also you should say how far it is roughly and which side of the street it is on. "It's about two minutes' walk from the station." "It's on the right-hand side." Giving directions takes practices but becomes easier over time.

你可用「Turn left.」（左轉）或「Turn right.」（右轉）、「Go straight.」（直走）、「It's the building next to the police station.」（就是警察局旁邊那棟。）等說法。另外也可描述一下大致距離或位於街道的哪一側等，例如：「It's about two minutes' walk from the station.」（從車站走過來約 2 分鐘。）、「It's on the right-hand side.」（在右手邊）。指引方向是需要練習的，一回生兩回熟。

許多人都說，美國大學「入學容易畢業難」。大部分的學校都沒有開學典禮，但是畢業典禮則會盛大慶祝。大學畢業可說是整個家族的喜事。

Subject: Good News!

Dear Doug and Steve,

[1]Just wanted to share some good news. [2]I'm graduating from college on May 7th! [3]I'm planning a little get-together after the ceremony at my parents' house. [4]You're welcome to drop in if you have time. [5]Hope you can make it!

Hanako

翻譯

主旨：好消息！

親愛的 Doug 與 Steve：
在此想與你們分享一個好消息。我 5 月 7 日就要從大學畢業了！我計畫於畢業典禮後，在我爸媽家辦個小聚會。如果你們有空，很歡迎你們過來。希望你們會到！

花子

 重要片語 & 句型

■ **Just wanted to share some good news.** 在此想與你們分享一些好消息。
■ **get-together** 聚會
■ **You're welcome to drop in.** 很歡迎你們過來。
■ **if you have time** 如果你們有空
■ **Hope you can make it!** 希望你們會到！（Note：省略了主詞 I）

佳句便利貼

■ Let me tell you my good news!
讓我告訴你我的好消息！

■ I've got great news to share.
我有好消息要分享。

■ I'm having a little party tonight if you're available.
我今天晚上要辦一個小型派對，如果你有空可以來參加。

■ There's a little soiree at Michael's apartment tonight.
今晚在 Michael 的公寓裡有個小聚會。

■ After the event is done, we'll all be going out for dinner.
活動結束後，我們大家會一起去吃晚飯。

■ Stop over when you have some time to chat.
你有空聊一聊的時候就過來。

 FAQ 93 敲定生意的關鍵

Question
How can I express shock or disbelief in an email message?

如何在電子郵件中表達震驚或難以置信？

Answer
"What?!," "I can't believe it.," "That's surprising." are all ways to show disbelief in an email message. The use of the question mark and exclamation point together show this very well.

「What?!」、「I can't believe it.」、「That's surprising.」等句子都可用於 email 中以表達難以置信。如果同時使用問號與驚嘆號，更能有效傳達這種感覺。

升遷通知
Announcing One's Promotion

許多人對於自行通知大家自己升官了的做法會有所遲疑，覺得反正別人總有一天會知道，所以沒必要特地通知大家。不過，以下範例則將升官的喜悅充分地傳達了出去。

Subject: Jack-in-the-Box's newest manager!

To all my friends and family,

[1]I got promoted! [2]From next month, you can call me Manager! [3]After five years at Jack-in-the-Box, I finally will be the one in charge of my own restaurant. [4]I hope you will all come on down to the restaurant and make me look good!

Tomohiro

翻譯

主旨：Jack-in-the-Box 新任店經理！

給我所有的朋友及家人：
我升官了！從下個月起，你們可以叫我店經理囉！在 Jack-in-the-Box 工作了 5 年，我終於能負責管理自己的餐廳了。希望你們大家都來餐廳替我捧捧場！

智弘

 重要片語 & 句型

■ **I got promoted!** 我升官了！
■ **You can call me ...!** 你可以叫我～囉！
■ **in charge of ...** 負責～
■ **come on down to ...** 過來～（= come over to）
■ **make** *someone* **look good** 替某人捧場

216

佳句便利貼

- I'm up for a promotion.
  我被提請升遷。

- My promotion got approved.
  我獲准升遷了。

- I did it! I'm the newest head of research at Kline Technologies.
  我做到了！我現在是 Kline 科技公司研究部門的新主管。

- You are looking at the new CEO of Bolt Paper.
  你眼前的就是 Bolt 紙業的新執行長。

- I got passed over for a promotion.
  我沒能成功升遷。

 **FAQ 94** 敲定生意的關鍵

Question

What are some more conversational ways to say "How are you?" in English?

英文中有沒有一些更口語的方式可表達「How are you?」之意？

Answer

"How's it going?" or "What's up?" are very common ways to sound more conversational and both sound friendly.

「How's it going?」或「What's up?」都是常聽到的說法，不僅更口語化，聽起來也都很友善。

新居落成通知
Announcing a New Home

我們有時會以 email 通知親朋好友新居落成,並順便邀請大家來參觀。就像買了新車便想開出來炫耀一樣,有了新房子也一定會想讓大家看看,這樣的想法似乎舉世皆然。

Subject: Construction update

To all our friends and family,

[1]We are happy to announce that construction is done on our new house and we are in the process of getting moved in. [2]We'll keep you informed of how things are going and, once we get situated, we'll start inviting people over for a visit. [3]Best wishes to all of you.

Kotaro and Amin Sato

翻譯

主旨:房屋建造工程最新消息

給我們所有的朋友和家人:
很高興在此宣布,我們的新房子已完工,而我們正在準備搬家中。我們會持續通知各位最新進展。一旦安頓好之後,我們就會開始邀請大家過來參觀。祝福大家。

佐藤幸太郎與亞美

 重要片語 & 句型

■ **We are in the process of getting moved in.** 我們正在準備搬家中。
■ **We'll keep you informed of** ... 我們會持續通知各位有關~的訊息。
■ **get situated** 安頓好
■ **Best wishes to all of you.** 祝福大家。

佳句便利貼

■ The work is complete on our new two-story house.
我們的兩層樓新屋已完工。

■ The building is finished and now we are looking for tenants.
大樓已完工，現在正在找承租人。

■ When we get settled, you're welcome to drop by for a visit.
我們安頓好之後，歡迎你們過來參觀。

 FAQ 95　敲定生意的關鍵

Question
How can I ask for someone's advice in English?

該如何用英文請他人提供意見？

Answer
"I need your opinion on something." or "What do you think I should do?" or "What would you do in this situation?" are three ways to request someone's opinion.

比方說「I need your opinion on something.」（我想聽聽你對某事的意見。）、「What do you think I should do?」（你覺得我應該怎麼做？），或「What would you do in this situation?」（在這情況下你會怎麼做？）這三句話都是徵詢對方意見的說法。

以下範例與其說是結婚通知，其實更像是邀請對方參加婚禮用的喜帖。結婚時負責聯繫的往往不是當事人，而是由家人或朋友代勞。下例中除了新郎、新娘的簡介外，也包含了婚禮地點說明及日期等資訊。

Subject: Engagement (RSVP requested)

[1]We are pleased to announce the engagement of our son Takayuki Yamamoto to Miss Katrina Juliette Watson. [2]The bride is the daughter of Dan and Priscilla Watson of Fairfield, Michigan. [3]The couple will be married in a service at the Church by the Side of the Road in Rockton on February 14th at 3 p.m. [4]You are cordially invited to attend. [5]A gift registry has been arranged at Nieman-Marcus.

Hiroki and Mie Yamamoto

（翻譯）

主旨：訂婚通知（敬請回覆是否參加）

在此很高興宣布，我們的兒子山本隆之與 Katrina Juliette Watson 小姐訂婚了。新娘是 Dan 和 Priscilla Watson 的女兒，來自密西根州費爾菲爾德。這對新人將於 2 月 14 日下午 3 點，在羅克頓的 By the Side of the Road 教堂舉行婚禮。誠摯地邀請您前來參加。賀禮清單已登錄於 Nieman-Marcus 百貨公司。

山本弘樹與美惠

 重要片語 & 句型

- **engagement** 婚約；訂婚
- **RSVP requested.** 敬請回覆是否參加。
- **be married in a service** 舉行婚禮
- **You are cordially invited to attend.** 誠摯地邀請您前來參加。
- **gift registry** 賀禮清單（在西方，為了避免收到重複的東西，新婚夫妻會於某間商店或百貨公司事先挑選出一些他們希望收到的賀禮，並登錄於店家的電腦系統中，以便親友認購。）

佳句便利貼

■ We are pleased to announce the union of Jim and Jeanie.
我們很高興宣布 Jim 與 Jeannie 結為夫妻。

■ We got engaged!
我們訂婚了！

■ Louise and Maurice got married in a civil service at the courthouse.
Louise 和 Maurice 已在法院公證結婚。（在英美等國，不想於教堂等宗教單位結婚的新人也可選擇到法院公證結婚。）

■ The couple got married in a private ceremony.
這對情侶在一場私人婚禮中結為夫妻。

■ BYOB requested.
請自備酒類飲料。

■ Volunteers requested.
招募（在婚禮上幫忙的）義工

 **FAQ 96** 敲定生意的關鍵

Question
What does RSVP mean in an invitation?　　邀請函裡的 RSVP 代表什麼意思？

Answer
RSVP is originally from the French words "Répondez s'il vous plaît" which means "Please let me know if you can make it to the event or not."

RSVP 這個縮寫源自於法文的「Répondez s'il vous plaît」，意思是「請讓我知道您是否能參加」。

本例為通知寶寶出生的 email。以下這封信是由因太太生了雙胞胎而興奮不已的丈夫所送出，其中提到了寶寶的體重，同時也傳達了母子均安的訊息。

Subject: Twins!

[1]It's a boy! [2]It's a girl!

[3]We are happy to announce the birth of our twins Evan and Samantha on November 20th. [4]Each baby weighed nearly 2,600 grams at birth. [5]Both mother and babies are doing great. [6]We hope you will share in this joyous moment with us in welcoming these two little people into the world.

The happy parents

翻譯

主旨：雙胞胎！

是男孩！還有女孩！
很高興通知各位，我們的雙胞胎 Evan 與 Samantha 已於 11 月 20 日出生。兩個寶寶出生時體重都接近 2,600 公克。母子均安。希望您能與我們分享這喜悅的一刻，一起歡迎這兩位小朋友來到世上。

幸福的爸媽

 重要片語 & 句型

■ **We are happy to announce the birth of** ... 很高興通知各位，～已出生。
■ **weigh nearly** ... 體重接近～
■ **at birth** 出生時
■ **Both mother and child are doing great.** 母子均安。
■ **We hope you will share in this joyous moment with us in welcoming** ... **into the world.** 希望您能與我們分享這喜悅的一刻，一起歡迎～來到世上。

佳句便利貼

■ You're going to have triplets!
你懷的是三胞胎！

■ My little girl weighs approximately 27kg.
我的小女兒體重約有 27 公斤。

■ When he was born, my son had lots of hair.
出生的時候，我兒子頭髮很多。

 FAQ 97 敲定生意的關鍵

Question

What are some ways to make a suggestion in English?

用英文提出建議有哪些方式？

Answer

You might want to say "How about …?" or "Why don't we …?," "I recommend …" For example, "How about going to a movie tonight with me?" "Why don't we see that new comedy that's playing?" "I recommend the lasagna at this restaurant. It's wonderful!"

你可以用「How about... ?」、「Why don't we... ?」或者是「I recommend ...」等句型來表達，例如：「How about going to a movie tonight with me?」（今晚跟我一起去看電影如何？）、「Why don't we go see that new comedy that's playing?」（我們何不去看那部正在上映的喜劇？）、「I recommend the lasagna at this restaurant. It's wonderful!」（我推薦這家餐廳的千層麵。超美味！）

98 出院通知
Announcing Getting out of the Hospital

基於禮貌，出院時應該通知住院時曾來探望，或曾寄 email、卡片來慰問的人。
這類 email 的起頭部分可直接使用下例寫法。

Subject: Out of the hospital

To everyone who has been so kind to me recently,

[1]I'd like to thank each and every one of you for the kindness and concern shown to me during my stay in the hospital. [2]I came through surgery just fine and was released from the hospital this morning. [3]It's great to be back and I feel healthier than ever. [4]I appreciate all of you.

Kazuhisa Nagasawa

翻譯

主旨：出院了

給每一位近來對我特別關切的朋友們：
在此要感謝你們每一位在住院期間對我表達的好意與關懷。手術順利完成，我今早已出院。回來真好，我覺得自己變得比以往更健康了。感謝大家。

長澤和久

 重要片語 & 句型

■ **To everyone who has been so kind to me recently,**
給每一位近來對我特別關切的朋友們：

■ **I'd like to thank each and every one of you for the kindness and concern shown to me during my stay in the hospital.**
在此要感謝你們每一位在住院期間對我表達的好意與關懷。

■ **come through surgery** 安然度過手術

■ **be released from the hospital** 出院

■ **It's great to be back.** 回來真好。

■ **I feel healthier than ever.** 我覺得自己變得比以往更健康了。

■ **I appreciate all of you.** 感謝大家。

佳句便利貼

■ I wish to thank all of you for your kind words to me.
我要感謝你們大家對我的慰問。

■ The operation on my elbow was successful.
我的手肘手術很成功。

■ I'm feeling better than ever.
我現在感覺非常好。

 **FAQ 98** 敲定生意的關鍵

Question

How can I give advice to someone in English?　該如何用英文給人忠告？

Answer

Try using "I think you should ..." or "If I were you, I would ..." as in the following. "I think you should ask her what she really thinks." or "If I were you, I would stop drinking so much."

你可嘗試以「I think you should ...」或者是「If I were you, I would ...」等句型來表達。例如以下兩個句子：「I think you should ask her what she really thinks.」（我覺得你該問問她真正的想法是什麼。）或「If I were you, I would stop drinking so much.」（假如我是你的話，就不會再喝這麼多酒了。）

附上照片
Attaching Pictures

寄 email 時如果有附照片，往往也得加上照片說明。例如，可以告訴對方這是在何時、何地拍攝的照片。另外，針對初次寄送照片的對象，還應進一步確認對方是否看得到附件圖檔。

Subject: Trip photos

Tasha,

[1]Did you recover from the trip? [2]Along with this message, I'm attaching several pictures that I took of you and me on our trip. [3]I hope you can see them. [4]If not, let me know. [5]Enjoy!

Izumi

翻譯

主旨：旅遊照片

Tasha：

旅遊的疲憊已經恢復了嗎？我在此信中附加了幾張我在此次旅遊中幫妳和我所拍攝的照片。希望妳看得到這些照片。如果看不到，告訴我一聲。好好欣賞吧！

泉美

重要片語 & 句型

■ **Did you recover from the trip?** 旅遊的疲憊已經恢復了嗎？
■ **Along with this message, I'm attaching ...** 我在此信中附加了～。
■ **If not, let me know.** 如果不，請通知我一聲。
■ **Enjoy!** 好好欣賞吧！

佳句便利貼

■ I'm sending the documents as an attachment to this message.
我將文件以附件形式附在此封信中寄出。

■ Attached to this message is the file you requested.
此信之附件即為你所要求的檔案。

■ With this message, I got a file I couldn't open.
此信的附加檔案我無法開啟。

 **FAQ 99** 敲定生意的關鍵

Question

I heard that an invitation from an American like "Stop by for dinner sometime." is not an invitation at all. How can we tell if it is a real invitation or a false one?

我聽說當美國人說「Stop by for dinner sometime.」（有機會請過來吃個晚飯。）時，根本不是真的要邀請你。到底該如何判斷哪個邀請是真的，哪個邀請是假的？

Answer

It is true that Americans sometimes make invitations they don't really mean to keep. The clue to a real invitation, however, is that the person inviting you will give an actual date and time. "Can you come for dinner on Friday? We'll start at 7 at Maruta." It's sometimes difficult to tell the difference, but over time you will come to understand which are real and which are not.

美國人有時確實會提出一些言不由衷的邀請。可用來判斷的線索就是，當對方真的想邀請你時，應該會提出明確的日期與時間，如「Can you come for dinner on Friday? We'll start at 7 at Maruta.」（你週五晚上可以來吃頓晚飯嗎？我們 7 點會在 Maruta 餐廳準時開始。）真假的差異有時很難判別，但是時間久了之後，你就會知道哪種邀約是真的，哪種又是假的了。

對所附照片的回應
Responding to a Message with Pictures

本例是針對對方寄來照片回覆感謝之意的 email。在簡短的郵件中，基本上應傳達 3 個重點：1. 附件照片可開啓；2. 對照片的感想；3. 隨照片想起之回憶或其他事情。

Subject: Thanks for the photos!

Anne,

[1]They look great! [2]Can you believe it has already been a month since we went? [3]Time flies! [4]Anyway, thanks again. [5]Hope we can get together again soon.

Tony

翻譯

主旨：謝謝你的照片！

Anne：

照片看起來真棒！妳相信嗎？從我們去旅遊到現在都已經過了一個月了！真是時光飛逝！總之，再次感謝妳。希望我們很快又能聚在一起。

Tony

 重要片語 & 句型

■ **They look great!** 它們看起來真棒！
■ **Can you believe it has already been ... since ...?**
　你相信嗎？從～到現在都已經過了～！
■ **Time flies!** 時光飛逝！
■ **anyway** 總之；無論如何
■ **Hope we can get together again soon.** 希望我們很快又能聚在一起。

佳句便利貼

■ Isn't it hard to believe we met 22 years ago?
實在很難相信我們已認識 22 年了吧？

■ Has it really been five years since we last met?
從我們上次見面至今真的已過了 5 年嗎？

■ Time passes quickly.
時間過得很快。

■ Anyhow, thanks for all your help.
無論如何，感謝您的幫忙。

■ Let's meet next time you're in town.
下次你到這裡來時，咱們見個面吧。

■ I hope we can meet up sometime after the holidays.
希望假期過後我們有機會碰個頭。

 FAQ 100 敲定生意的關鍵

Question

What are some other ways to say good-bye, besides "Good-bye.?"

除了「Good-bye.」之外，英文還有哪些道別的說法？

Answer

"Talk to you later." "See you later." "Have a good day." All of these function in the same way as good-bye, and they sound much more natural.

「Talk to you later.」、「See you later.」、「Have a good day.」這些說法都具備和 good-bye 一樣的功能，而且聽起來更自然。

Q&A

隨查即用！
商英 Email 好用句 400

✉ 下列英文當中，粗體字表示為重要句型，非粗體字部分則可視實際狀況
　　替換其他詞彙，以便因應各種需求。

開頭敬稱

☐ **Dear Mr. John Perry,**	John Perry 先生，您好：
☐ **Dear Ms. Perry,**	Perry 小姐，您好：
☐ **Dear Sir or Madam:**	各位先生女士：
☐ **To whom it may concern:**	敬啓者：
☐ **Hello,**	您好：
☐ **Hi,**	嗨：
☐ **Greetings!**	您好！
☐ **John,**	John：
☐ **Everyone,**	各位：
☐ **Dear friends,**	親愛的朋友們：
☐ **All Staff,**	各位同仁：
☐ **To all employees,**	致全體員工：
☐ **To all staff,**	致所有員工：
☐ **To all interested employees,**	致所有有興趣的同仁們：
☐ **To the employees of Teshio Industries,**	致天鹽工業全體員工：
☐ **To all our loyal customers,**	致我們所有的忠實顧客們：
☐ **To all our valued customers,**	致我們所有尊貴的顧客們：
☐ **To all our valued members,**	致所有敬重的夥伴們：
☐ **To all Value Fund shareholders,**	致各位 Value Fund 的股東們：
☐ **To all my [our] friends and family,**	給我（們）所有的朋友及家人：
☐ **To everyone who has been so kind to me recently,**	給每一位近來對我特別關切的朋友們：
☐ **To the Allens,**	給 Allen 全家：
☐ **To the Harris family,**	給 Harris 家族：
☐ **To Tom, Sylvia and the kids,**	給 Tom、Sylvia 以及孩子們：

信尾敬詞

☐ **Sincerely (yours),**	敬啓

Best [Kindest] wishes,	祝好
Best [Kindest / Warmest] regards,	祝好
Regards,	祝好
All the best,	祝福您
Thank you,	謝謝
Thank you again,	再次感謝
Thank you in advance,	先謝了
Thanks,	謝謝
With gratitude,	萬分感激
With greatest respect and admiration,	謹致上最崇高敬意
Best wishes to all of you.	祝福大家。
Have a great day!	祝你有個美好的一天！
Looking forward to hearing from you,	期待您的回覆。
Looking forward to seeing you.	期待與您見面。

起頭句

My name is Kumiko Tanimoto **and I'm responsible for** sales at Muroran Steel.	我是負責 Muroran 鋼鐵公司業務工作的谷本久美子。
My name is Naoko Yoshida, **and I'm** the Human Resources chief.	我是吉田直子，人力資源部的主任。
This is Saeko Shimada **over at** ABC Supply.	我是 ABC 器材的島田紗枝子。
I'm writing to you today about ABC Industries' new 7800C water pump.	今日寫這封信給您的目的是想詢問有關 ABC 工業的新 7800C 水泵產品。
I'm writing to inquire about your latest product.	我來信的目的是想詢問貴公司的最新產品資訊。
I'm writing to let you know the results of last week's business trip to our offices in Frankfurt.	在此寫信向您報告我上週到法蘭克福分公司出差的結果。
I'm writing to let you know that our order hasn't arrived.	寫這封信的目的是要通知您，我們訂購的商品尚未送達。
I'm writing to complain about the service at your restaurant this evening.	我寫此信的目的是要投訴貴餐廳今晚的服務。
I'm just checking to see what forms of payment you accept.	我想確認一下貴公司接受哪些付款方式。

Just want to let you know I've made the payment.

謹通知您一聲我已付款。

Just wanted to share some good news.

在此想與你們分享一些好消息。

Just following up to let you know that I deposited the money for the office furniture into your corporate account this afternoon.

謹在此通知您，今天下午我已將辦公室家具的款項存入貴公司帳戶。

I have an announcement to make.

我有件事要宣布。

I've got great news to share.

我有好消息要分享。

Let me tell you my good news!

讓我告訴你我的好消息！

Three Marks Industries **is pleased to announce** Shin-ichi Saito has been selected our new company president.

三標工業很高興宣布，齋藤新一已獲選為本公司新總裁。

We are happy to announce that construction is done on our new house and we are in the process of getting moved in.

很高興在此宣布，我們的新房子已完工，而我們正在準備搬家中。

This is just a heads-up to let you know that your order is on the truck.

本信的目的是要通知您注意，您所訂購的商品已經裝車出貨。

This is to inform you that your rent is overdue.

在此通知您，您的房租逾期未繳。

This is to notify you that your account as of July 31st has become delinquent to the amount of 432.76.

在此通知您，截至 7 月 31 日為止，您的帳戶已拖欠 432.76 美元。

This message is to acknowledge receipt of the parts shipment order #84123.

本信的目的是要通知您，我們已收到訂單編號 #84123 之零件。

I'm sorry to inform you that I won't be able to make today.

很抱歉必須通知您，我今天沒辦法到公司上班。

I regret to inform you that the office chair you ordered is currently out of stock.

很遺憾必須通知您，您所訂購的辦公椅目前缺貨中。

附件

Enclosed please find the bill for the repair to your office window.

隨信附上貴辦公室窗戶修繕之帳單明細。

Attached to this memo is an invoice.

隨本信附上發票

Attached to this message is a breakdown of the costs.

隨信附上費用明細。

Here are the receipts and my expense report from my trip to India.

這些是我去印度出差時的收據和支出報告。

| **Here is** the payment request for services rendered. | 這是為您服務的費用請款單。 |
| **I'm sending** the document **as an attachment to this message.** | 我將文件以附件形式附在此封信中寄出。 |

維繫關係

Thank you for your business.	謝謝您的惠顧。
We appreciate your business.	感謝您的惠顧。
Thank you and we appreciate your business.	非常感謝您的惠顧。
It is a pleasure doing business with you.	很高興能與您合作。
From all of us at Brio Plastics **we thank you for your business.**	Brio 塑膠公司全體員工感謝您的惠顧。
Thanks and we look forward to working with you.	感謝您，同時很期待與貴公司合作。
Thanks and we look forward to serving you again in the future.	謝謝，也期待未來能再次為您服務。
We hope to serve you again soon.	希望很快能再次為您服務。
We are certain to do business with you again.	我們一定會再次合作的。
I hope that we will have the opportunity to do business together now and in the future.	誠摯希望從今而後能有機會與您合作。
It would be a pleasure to work together again in the future.	很期待未來還有機會合作。
We look forward to doing business with you again in the future.	期待未來能再次與您合作。
We welcome the opportunity to work with you in the future.	我們很期待將來有機會與您合作。

請託

Please let me know.	請通知我一聲。
Please let me know if you can join us.	請告訴我你是否能夠加入我們的行列。
Anyway, let me know.	無論如何，請通知我一聲。
If this works for you, please let me know.	如果您覺得這樣可行，就請通知我一聲。
Your prompt attention in this matter would be appreciated.	如果您能儘速處理此事，我將非常感激。

Would you take a moment to answer a few questions?	能否請您花點時間回答幾個問題？
Please respond at your earliest convenience.	請儘快回覆。
Could you check on that for me?	這部分可以幫我確認一下嗎？
If you could check on that and get back to me, I would appreciate it.	如果您能確認一下並作回覆，我將萬分感激。
If you have any information, could you please let me know?	如果您有任何訊息，能否讓我知道？
I'd like some more information about [on] your products.	我希望能取得更多貴公司的產品資訊。
Any information you could provide would be greatly appreciated.	只要你們提供任何資訊，我都會非常感激。
Please forward the information to the address below, care of me.	請將資訊寄至下列地址，收件人寫我即可。
Please forward the information, in care of Dr. Ami Watanuki.	請將資訊寄給貴亞美博士，由她轉交。
Could you send me some information about [on] your fall line?	能否請您寄一些貴公司的秋季系列商品資訊給我？
Could I get a brochure about your company's products?	能否提供貴公司產品型錄給我？
I was wondering if you could send me a brochure about your new product.	能否請您寄一份貴公司的新產品型錄過來？
Would you mind giving us an update on the upgrades to our computer system?	能否請您提供我們電腦系統升級作業的最新狀況？
I was wondering if you could have one of your sales team stop over with the information.	能否請貴公司派一位銷售人員前來提供資訊？
Please complete these documents and give them back to me.	請填妥這些文件後再寄回給我。
Please send it directly to me.	請直接寄給我本人。
Please send me the address and any other information that might be useful.	請將地址與任何其他可能用得到的資訊都寄給我。
Please understand this fact.	這件事還請見諒。
If you are able to participate, your expertise would be a great help.	如果你能加入，你的專業知識將成為莫大助益。

提案

I'd like to make a proposal.	我想提個案。
Allow me to make a suggestion.	容我提出一個建議。
Here's an idea!	我有個好點子！
I have a great idea.	我有個很棒的點子。
I have this idea I've been working on.	我一直有個想法。
Why don't we consider this idea?	我們何不考慮採用這個構想？
I'd like to run it by you.	希望能說給您聽聽，看看您有什麼意見。
How about if we offer flex time to all employees**?**	您覺得對所有員工實施彈性工時制如何？
Wouldn't it make sense to try something new**?**	試試新東西不是比較明智嗎？
What a great idea!	這點子真棒！
That's not a very good idea.	這構想不太好。
What if we do this instead?	如果改成這樣做呢？
We welcome all suggestions.	我們歡迎所有建議。
Rest assured, your suggestion will be considered carefully.	放心，我們一定會仔細考慮您的建議。
You can be assured that your suggestion will receive careful consideration.	請放心，我們一定會仔細考慮您的建議。
You can be certain that your suggestion will be taken seriously.	您可以放心，我們一定會慎重考慮您的建議。

提供協助

If I can be of any more help, please let me know.	如果還有什麼能幫得上忙的，就請通知我一聲。
If I can be of further assistance, get in touch with me again.	如果還有什麼我幫得上忙的，請再次跟我聯絡。
If I can be of further help, feel free to let me know.	如果還有其他需要幫忙的，請告訴我一聲，不用客氣。
If I could offer any further information, please do not hesitate to contact me.	如果您需要任何進一步資訊，請不吝與我聯繫。
If we can of any assistance, don't hesitate to give us a call.	如果有什麼我們幫得上忙的，別客氣，請來電告知。
If we can be of further assistance, please feel free to contact me directly.	如果還有什麼我們能幫得上忙的，別客氣，請直接與我聯繫。

If you need any other information about the trip, **please let me know.**	如果您還需要任何其他有關此次出差之資訊，請通知我一聲。
If you need anything, give me a call at 453-7128.	如果您有任何需要，請撥 453-7128 這支電話給我。
If you need further details, please let me know.	如果您需要更多詳細資料，請通知我一聲。
If you need past order history, **please contact** Lou in accounting.	如果您需要以往的訂購紀錄，請與會計部門的 Lou 聯繫。
In the event that you need anything, please feel free to contact me.	要是您需要什麼，請別客氣，就跟我聯絡。
Please contact me if you need any further assistance.	如果您需要進一步協助，請與我聯繫。
If there are any problems, let me know.	如果有任何問題，請通知我一聲。
Any problems, give me a call.	如果有任何問題，請撥電話給我。
Any questions may be directed to me.	如果有任何疑問，請直接與我聯繫。
Questions about the program may be forwarded to my secretary.	關於此計畫的所有問題都請傳給我的秘書。
Direct all questions to our lawyers.	所有問題都請直接向本公司律師（法律顧問）提出。
Give me a call if you need anything.	如果有任何需要，請撥電話給我。
Don't think twice about getting help if you need it.	如果您需要幫忙，請別猶豫，立刻尋求援手。
Please feel free to call me at this number.	請別客氣，就打這支電話與我聯繫。
Should you need anything, call this number, 423-8761.	萬一您需要什麼東西，請電 423-8761。
We will deal with this situation later.	我們稍後會處理這個狀況。
Here is all the information I have on the current Bakersfield order.	這些是我手上所有和 Bakersfield 公司目前訂單相關的資訊。
I have no other knowledge except the information I have already given you.	除了已給您的資訊外，其他事情我並不清楚。

會議

We'd like to have a meeting on Wednesday **at** 1 **in** the conference room.	我們想在週三下午 1 點於會議室開個會。

☐	**There will be a meeting** this Wednesday **at** 10:30 a.m. **in** the staff room.	本週三上午 10 點 30 分我們將在員工休息室舉行會議。
☐	**We'll be meeting in** Meeting Room 4.	我們將在第 4 會議室見面。
☐	**There is a mandatory meeting at** 5.	5 點有個強制參加的會議。
☐	**There will be an informal brainstorming meeting to drum up ideas for** our fall promotional campaign.	我們將舉辦一場非正式的腦力激盪會議，來為我們的秋季促銷活動募集新點子。
☐	**The meeting is about** upcoming outsourcing.	此會議討論的是即將進行之外包作業。
☐	**The topic of the meeting will be** our current budget crisis.	會議主題是關於我們目前所面臨之預算危機。
☐	**The agenda for the meeting is still undecided.**	會議議題仍未定。
☐	**This calls for a meeting.**	這需要開會討論。
☐	**Let's discuss this over** dinner.	讓我們邊吃晚飯邊談吧。
☐	**Let's set up a working** lunch.	讓我們訂個午餐會吧。
☐	**We will delay making a decision until** after next week's meeting.	我們將延至下週會議後再作決定。
☐	**All members, except those away on company business, are expected to attend.**	除了出差的同仁外，所有員工都應出席。
☐	**If for some reason you cannot attend, please contact your supervisor.**	如果有特殊原因無法參加，請與您的主管聯繫。
☐	**Attendance is not mandatory, but we would like the input of as many people as possible.**	此聚會並不強制出席，但是我們仍希望大家都能踴躍提供創意。
☐	**Attendance is optional at the meeting.**	此會議可自由出席。
☐	**I can't make it to the meeting on** Friday.	我沒辦法參加週五的會議。
☐	**I'm afraid I'm unable to attend the meeting due to** a family emergency.	因為家中有急事，我恐怕無法出席會議。
☐	**I have no choice but to cancel the meeting.**	我不得不取消會議。
☐	**Ask your superior if you have any questions.**	如果有問題，請洽詢您的主管。
☐	**Several new ideas came up in the meeting that you may want to think about.**	此次會議中有幾個新點子，您或許會想考慮看看。

Effective immediately, the use of social networking sites as Friendbook and Twizzer during company time **is prohibited.**

上班時間禁止使用 Friendbook 和 Twizzer 之類的社群網站，本規定即刻生效。

Our new program **will go into effect in** August.

我們的新計畫將於 8 月開始實施。

Starting today, the company will no longer pay for employee parking.

從今天起，公司不再支付員工的停車費。

This policy is effective immediately.

此項政策立即生效。

This will be effective from September 1st.

此事項將於 9 月 1 日起生效。

This is company time.

現在是上班時間。

No personal calls on company time, please.

上班時間請不要講私人電話。

Employees will be warned if they are breaking company rules.

員工如果違反公司規定，就會被警告。

This is your final warning.

這是最後一次警告。

I would like to bring to your attention the memo I received from our corporate headquarters.

在此要通知各位一則我從總公司收到的備忘錄。

Please be aware that our company servers will be down for maintenance from 9 p.m. Friday, the 7th to 6 a.m. on Saturday, the 8th.

請注意，本公司伺服器因實施維修工程，將從 7 號週五晚上 9 點起至 8 號週六上午 6 點止暫停運作。

Please take notice that offices will be closed for the Easter holiday.

請注意，復活節假期期間辦公室將會關閉。

We will be closed for a holiday from March 13th **through** 31st.

從 3 月 13 至 31 日為止，我們將因放假而暫停營業。

Our offices will close on December 25th **for** the Christmas holiday.

12 月 25 日本辦公室將因聖誕節假日而關閉一天。

All order received by midnight on March 10th **will be processed ahead of the vacation.**

所有在 3 月 10 日午夜前收到的訂單，都將於假期前處理完畢。

All orders received after March 10th **will be processed beginning** April 1st.

3 月 11 日以後收到的訂單，將於 4 月 1 日起開始處理。

I would like to file a complaint.

我要提出客訴。

I would like to express my dissatisfaction with your poor service.

我想針對貴公司的劣質服務表達不滿。

I'm very unhappy about the quality of your work.	我對貴公司的工作品質非常不滿意。
Your service is not up to par.	貴公司的服務根本不及格。
What I received was incorrect.	我收到的商品不對。
Please inform me of the status of my package.	請告知我包裹目前配送狀況。
As of the start of business this morning we still do not have our package.	直至今早上班時，我們的包裹仍未送到。
As of this morning, I haven't heard anything from the head office.	截至今日上午為止，我還沒收到總公司傳來的任何消息。
Could you fix this mistake?	能否請你們更正這錯誤？
Could you please remedy this situation?	能否請你們針對這情況做點補救？
We appreciate your business and sincerely regret that your package did not arrive on time.	感謝您的惠顧，同時也為您的包裹未能準時送達一事表達誠摯歉意。
We are surprised that our services did not match your expectations.	很訝異我們的服務未能符合您期待。
It seems that there is a problem somewhere in our supply chain.	看來我們供應鏈中的某個環節是有問題的。
Obviously, mistakes have been made.	顯然錯誤已經造成。
Clearly this time we failed.	很顯然這次我們失敗了。
All comments pertaining to our service will be directed to the appropriate department.	所有關於敝公司服務的意見都會傳達給相關部門。
Though weather delays are not in our control, we do regret the inconvenience it causes.	雖然天候導致的延遲並非我們所能掌控，但是因此造成不便，我們仍深感遺憾。
While this was beyond our control, we would like to offer you a voucher for free shipping on your next package.	雖然這個部分超出了我們的掌控範圍，但我們仍願意提供 1 張抵用券，您下次的包裹可免運送費。
While we make every effort to insure that all parts are inspected, occasionally a defective one escapes our best efforts.	雖然我們竭盡所能地確保所有零件都經過檢驗，但仍免不了百密一疏。
While we try our best, sometime we fail.	雖然我們竭盡所能，但仍偶有失誤。
Without a doubt, we need to raise the level of service to our customers.	毫無疑問地，我們需要提升客戶服務品質。
We will do our best to satisfy your business needs.	我們將盡最大努力來滿足您業務上的需求。

We work hard to bring you the finest foods from around the would.	我們竭力提供您來自世界各地的最佳美食。
I have spoken to the employees about your message and please be assured that we will continue to strive to improve our service.	我已向員工傳達了您的意見，請相信我們一定會繼續努力改善我們的服務品質。

催促

Thank you and your prompt response would be greatly appreciated.	謝謝。如果能儘快收到您的回覆，我將萬分感激。
Your prompt attention would be greatly appreciated.	如果您能儘速處理，我將非常感激。
Please let me know ASAP.	請儘快通知我。
As this is a priority, please get back to me ASAP.	這件事必須優先處理，所以請儘快回覆我。
As this is an urgent matter, please respond quickly.	這是急件，請趕快回覆。
Could you fedex the documents over **immediately?**	能否請您立刻將文件快遞過來？
Could you please ship our order **right away?**	能否請您立刻將我們訂購的商品出貨？
We need to finalize these contracts immediately.	我們必須立刻敲定這些合約。
Make this your priority.	請優先處理這件事。
Please pay immediately.	請立即付款。
We are desperate to get those parts.	我們非常需要那些零件。
We are desperate need of those replacements.	我們急需那些替換品。
We are in urgent need of color toner cartridges.	我們急需彩色碳粉匣。
We are up against a time limit.	我們的時間非常有限。
We're running out of time.	我們快沒時間了。
This is an urgent matter.	這是急件。
We have a strict deadline.	我們的期限緊迫。

Best wishes during this holiday season! 　在這佳節期間為您獻上最深祝福！

Happy Holidays! 　佳節愉快！

Season's Greetings! 　聖誕快樂！（用於賀卡上）

New year's wishes to your family! 　為您全家獻上新年祝福！

20XX will bring you much prosperity. 　祝您 20XX 年事業興旺。

Happy 20XX! 　20XX 年新年快樂！

Have a wonderful 20XX! 　祝您有個美好的 20XX 年！

I hope that 20XX brings you even more happiness and success! 　我希望 20XX 年能為您帶來更多快樂與成就！

I hope this year brings you everything you desire. 　祝您今年心想事成。

I hope this year welcomes you with much joy. 　祝您新的一年充滿歡樂。

I hope you will get everything your heart desires. 　祝您心想事成。

I hope you find everything you're looking for. 　祝您萬事如意。

Wishing you and your family a very merry Christmas and a happy new year! 　祝福您和全家聖誕快樂、新年快樂！

Here's hoping for the success of each of you. 　在此敬祝各位馬到成功。

Happy Valentine's Day, Sweetheart! 　親愛的，情人節快樂！

Happy Valentine's Day to the love of my life! 　情人節快樂，我的一生摯愛！

感謝

Thank you. 　謝謝您。

Thank you for your idea. 　感謝您提出您的構想。

Thanks for the photos of our trip to Europe. 　謝謝你寄來我們歐洲旅遊時的照片。

Thanks again. 　再次感謝。

This is a small token of my appreciation. 　這代表我的一點謝意。

We really appreciate it! 　我們真的十分感激！

I appreciate all of you. 　感謝大家。

Your kindness is greatly appreciated. 　非常感謝您的和善。

I really enjoyed working with you. 　與您共事真的很愉快。

It was a pleasure doing business with you.	很榮幸能與您合作。
Thank you for your prompt attention.	感謝您即時處理。
Thanks for asking for my help.	感謝您來尋求我的協助。
Thanks for thinking of me.	謝謝您想到了我。
Thanks for everything you did.	感謝您所做的一切。
Thanks for the heads-up!	感謝您的提醒！
Thanks for all your help.	感謝各位的協助。
Thanks for your request for my participation on your sales team.	謝謝您邀請我加入您的銷售團隊。
Thank you for your suggestion regarding the financial services division of our office.	感謝您對敝公司財務服務部門的相關建議。
I appreciate what you said about me to the promotion committee.	很感激您在晉升評核委員會中替我美言了幾句。
I appreciate your consideration.	非常感激您將我納入考慮；非常感激您的賞識。
We appreciate your understanding.	感謝您的體諒。
I'm grateful for your message asking for my assistance.	很感激您來信尋求我的援助。
We are grateful to you for paying us promptly.	感謝您即時付款。
We are thankful for your continued support of our fundraising campaign.	非常感謝您持續支持我們的募款活動。
Thanks to your help, we managed to increase our business with them an additional 14%.	由於您的幫助，我們能成功地將與該公司的交易量提升 14%。
We appreciate all input into our business and are grateful that you took the time to offer your assistance.	所有與敝公司業務相關之提案，我們都十分歡迎，也很感激您特地花時間提供協助。
We deeply appreciate your service to the company.	我們非常感謝您對公司的付出。
Our office appreciates all the hard work you did for us.	我們十分感激您對本公司所付出的努力。
We would like to thank you for the contributions you made to our office during your five years here in Tokyo.	感謝您在東京 5 年間對我們公司所做的貢獻。
It would be my pleasure to join the team.	我很樂意加入貴團隊。

We couldn't have done it without you.	若是沒有您,我們一定無法辦到。
We would have been lost without your help.	若是沒有您的協助,我們肯定已經失敗了。
Your comments about our service are greatly appreciated.	感謝您對敝公司服務所提出之相關意見。
Your understanding and cooperation on this matter will be greatly appreciated.	非常感謝您對此事的體諒與合作。
I'm happy to be considered by you for this important job.	很高興您考慮讓我擔任此重要工作。
Thank you for considering my feelings.	謝謝你顧慮到我的感受。
Thank you for keeping me in your thoughts while I am in the hospital.	謝謝你在我住院時仍掛念著我。
Thank you so much for the beautiful flowers you sent.	非常感謝你送來的美麗花朵。
Thank you so much for your kind words of encouragement about my transfer back to Japan.	非常感謝您因我即將轉調回日本,而送給我的鼓勵話語。
Thank you for your kindness and hospitality.	感謝您的和善與招待。
You are so thoughtful and considerate.	你真是體貼又周到。
I wish to thank all of you for your kind words to me.	我要感謝你們大家對我的慰問。
I'd like to thank each and every one of you for the kindness and concern shown tome during my stay in the hospital.	在此要感謝你們每一位在住院期間對我表達的好意與關懷。
Thank you for your transaction.	感謝您的交易。
Thanks again for your order.	再度感謝您的訂購。
We would like to thank you for your prompt shipment of our order.	感謝您迅速地將我們所訂的貨品送到。
You payment came in today and we thank you.	您支付的款項今日已入帳,非常感謝。
Thanks for the quick response.	感謝您迅速回覆。
Thank you for your quick response.	感謝您的迅速回應。
Thanks for getting back to me about the price of the used forklift.	謝謝您對中古堆高機價格的回覆。
Thanks for mailing me back.	感謝您的回信。

Thank you for sending me the catalog of your products.	感謝您寄來貴公司商品型錄。
Thanks you for quickly sending the required documents.	謝謝您迅速寄來所需文件。
Thank you for your request for an appointment.	感謝您來信預約會面。
I was an honor to meet you.	十分榮幸能與您見面。
I'm grateful for the opportunity to meet with you.	我很感激有機會能與您見面。
It was a pleasure speaking with you about our products.	很榮幸能與您談談敝公司產品。
I really enjoyed our conversation.	和您聊得真的很開心。
Thank you for offering to meet so that we might discuss our mutual business interests.	感謝您願意見面以便討論我們的互利業務。
Thank you once again for taking the time to meet with me last Friday.	再次感謝您上週五撥冗與我會面。

道歉

We apologize for this inconvenience.	很抱歉造成您的不便。
We apologize for this policy but it has become necessary.	很抱歉提出此規定，但目前確實有必要。
We sincerely apologize for neglecting to return your call.	對於忘了回電給您一事，我們深感抱歉。
I apologize for any trouble this may cause.	很抱歉造成您的困擾。
I apologize for this inconvenience.	很抱歉造成您的不便。
I apologize for the delay in making payment.	我為延遲付款一事道歉。
Please accept our apology.	請接受我們的歉意。
Out humble apologies for causing you trouble.	造成您的不便，我們深感抱歉。
Please allow me to apologize for the defective part you received from our company.	請容我針對您從敝公司收到的瑕疵品致歉。
My apologies.	很抱歉。
Please forgive this mistake.	請原諒這次的錯誤。
I'm sorry to ask you to repeat the process all over again.	很抱歉同一件事讓您重頭再做一遍。
I'm really sorry for being late in paying for the drywall work you did for us.	真的很抱歉拖欠您的砌牆工程款項。

246

I'm sorry I didn't pay you on time.	很抱歉沒能準時付錢給您。
We would like to express our deep regret that your package was not delivered on time.	對於您的包裹未能準時送達一事，我們深表遺憾。
We deeply regret the need to raise our prices.	真的很抱歉但我們必須漲價。

祝賀

Most sincere congratulations are in order.	獻上最誠摯的祝福。
Congratulations!	恭喜！
Congratulations on your big promotion to management!	恭喜你擢升管理階層！
Congratulations on your getting an MBA!	恭喜您取得 MBA 學位！
Congratulations on your rise to the top of the company.	恭賀您榮升公司最高階層。
Congrats on getting promoted!	恭喜你獲得升遷！
Congratulations on the start of your new business!	恭喜您開啟新事業！
Congratulations on your retirement!	恭喜您退休了！
Congratulations from the bottom of my heart.	從我心底獻上最深切的祝賀。
We were pleased to learn of your new business venture.	我們很高興聽到您要發展新事業。
I'm sure you will make an excellent manager.	我相信你一定會成為一位優秀的經理。
Your hard work has finally come to an end.	您辛苦的努力總算告一段落。
It's been a long time coming and you are very deserving.	真是期待已久，而你也當之無愧。
Many years of effort have led to this moment.	有多年努力才有這一刻。
You are worthy of every success that comes your way.	您的成功全都實至名歸。
You deserve it!	這是你應得的！
You will soon be rewarded for all that you have done in life.	您這一生中的所有付出即將得到回報。
You certainly have a bright future ahead of you!	你的前途肯定一片光明！
You must be excited.	你一定很高興吧。
Best wishes and continued happiness in this new chapter of your life.	在此獻上最大的祝福，願您在這人生新的一章中繼續過著幸福快樂的日子。

Good luck in your future!	祝你未來一路好運！
I wish you much success in your future endeavors!	祝你未來的努力都能成功！
I have enjoyed working with you over those past many years.	過去多年來與您一起工作非常愉快。
I hope you will have a wonderful retirement and that you can now do all those things you've dreamed of doing.	希望您能有美好的退休生活，並能做所有您夢想要做的事。
We got engaged!	我們訂婚了！
I was so happy when I got your announcement that you two got engaged!	得知你們兩位訂婚的消息時，我真的非常高興！
I heard you two are tying the knot.	我聽說你們倆打算結婚。
When is the wedding and where will it be held?	婚禮將於何時、何地舉行？
We are pleased to announce the union of Jim and Jeannie.	我們很高興宣布 Jim 與 Jeannie 結為夫妻。
I wish the future Mr. and Mrs. Smith a happy future together.	祝未來的 Smith 夫婦幸福美滿、永浴愛河。
I wish the soon-to-be newlyweds all the happiness in the world.	恭祝即將結合的新婚夫妻們幸福無限。
I wish you much happiness and joy together as you start your lives as husband and wife.	願你們的婚姻生活充滿幸福與喜悅。
I wish the happy couple a long life together.	願兩位佳偶天成、永浴愛河。
My heartfelt congratulations on your getting married!	誠摯地恭喜兩位新婚快樂！
Best wishes to you both!	在此為兩位獻上最深祝福！
Enjoy your new life as a married couple.	好好享受你們的新婚生活。
I'm so excited about your recent marriage to Trinity.	聽到你最近和 Trinity 結婚了，我非常高興。
You two are perfect together.	你們倆是天作之合。
You two were meant for each other!	你們倆真是天生一對！
May you be happy in your future.	願您未來幸福快樂。
May you find happiness.	願您找到幸福。
We are happy to announce the birth of our twins Jack and Rose on November 20th.	很高興通知各位，我們的雙胞胎 Jack 與 Rose 已於 11 月 20 日出生。

It's a boy!	是男孩！
It's a girl!	是女孩！
Both mother and the baby are doing great.	母子均安。
When he was born, my son had lots of hair.	出生的時候，我兒子頭髮很多。
We hope you will share in this joyous moment with us in welcoming these two little people into the world.	希望您能與我們分享這喜悅的一刻，一起歡迎這兩位小朋友來到世上。
Congratulations on the birth of your new baby girl / boy!	恭喜你們的女／男寶寶誕生！
I'm sure you'll both be the best parents a baby could ask for.	我相信你們一定會成為寶寶能擁有的最佳父母。
A child grows up quick, so cherish every moment!	孩子的成長非常迅速，所以請務必珍惜每一刻！
Enjoy every minute of your parenthood!	好好享受為人父母的每一刻！
Enjoy parenthood!	請好好享受育兒時光！
Happy birthday!	生日快樂！
I hope this year brings you everything you wish for and that there are many more birthdays to come!	祝你今年事事順心，並且福如東海、壽比南山！
Age brings experience.	歲月豐富了經歷。
I'm sure you are aging gracefully.	我確信隨著年齡增長，你變得愈來愈優雅。
The best is yet to come!	最棒的才剛要開始！
Your best years are ahead of you!	您意氣風發的年代即將來臨！
With age comes wisdom!	祝你添歲長智慧！
You are aging like a fine wine.	您就像瓶好酒，愈陳愈香。
You still look as young as ever!	您依舊青春如昔！
You've aged beautifully.	您成熟中更見美麗。

贈禮

I just sent a package over to you in celebration of your birthday.	為了祝賀你的生日，我剛剛寄了個包裹過去給你。
It's a little something I picked up on my recent trip to Tahiti.	那是我最近去大溪地旅遊時挑選的小東西。

It's something you've always wanted.	那是你一直都很想要的東西。
Here is the necklace you've been longing for.	這是你渴望已久的項鍊。
Here's a watch to mark this special occasion.	謹以此手錶來慶祝這個特別的時刻。
I found the book you've been wanting for a long time.	我找到了你一直想要的那本書。
I hope you like it.	希望你會喜歡。
I hope these flowers are to your liking.	希望你喜歡這些花。
I hope this vase suits your tastes.	希望這個花瓶你會喜歡。

傷病相關

James was hospitalized with pneumonia last night.	James 昨晚因肺炎入院。
Susumu was admitted to the hospital last night after complaining of chest pain.	Susumu 昨晚在抱怨胸痛後,就住進了醫院。
The operation on my elbow was successful.	我的手肘手術很成功。
I came through surgery just fine and was released from the hospital this morning.	手術順利完成,我今早已出院。
I'm feeling better than ever.	我現在感覺非常好。
It's great to be back and I feel healthier than ever.	能回來真好,我覺得自己變得比以往更健康了。
I was so sorry to hear that you are in the hospital.	很遺憾聽說您住院了。
I hope you'll make a quick recovery from your illness.	希望您的病能迅速痊癒。
I hope you'll be on the mend quickly and will be out of the hospital quickly.	希望您很快就可以康復出院。
Get well soon!	早日康復!
Best wishes!	祝福您!
All the best,	祝好

喪事弔唁

Sunstrand Industries is sad to report the loss of our beloved president Mr. Kenzaburo Miyagi.	Sunstrand 工業在此帶著悲痛通知您我們所敬愛的宮城健三郎總裁辭世的消息。

250

I'm sad to have to tell you about the death of my business partner.

很難過必須通知您我生意夥伴的死訊。

Did you hear about the passing of Mr. Misawa?

您有沒有聽說三澤先生過世的消息？

We were deeply saddened to learn of the loss of your valued president Mr. Kenzaburo Miyagi.

獲悉貴公司備受敬愛的總裁宮城健三郎先生過世之消息，我們深感悲痛。

I share your grief about the passing of Mr. Oldham.

對於 Oldham 先生過世，我與您同感悲傷。

My deepest sympathies about the loss of your beloved cat, Sasha.

對於你痛失愛貓 Sasha，我在此獻上最深切的哀悼之意。

My condolences on the loss of your friend.

在此為您痛失好友一事致上哀悼之意。

It was with great regret that we learned the news of her death late.

我們非常遺憾這麼晚才得知她過世的消息。

Mr. Miyagi **was a trusted partner and loyal friend to** our firm for many years.

多年來，宮城先生一直是位深受信賴的夥伴，也是我們忠實的好朋友。

Ron **was one of the best colleagues I ever worked with.**

Ron 是我所共事過最優秀的同事之一。

Ms. Honda **was a valuable asset to our company.**

本田小姐對本公司來說，是十分重要的人才。

He [She] will be missed.

我們會懷念他（她）的。

My dog **was a great partner in my life.**

我的狗兒曾是我生活中的絕佳夥伴。

A memorial service will be help on September 14th at 7p.m. **at** Yasuragi Funeral Home.

喪禮將於 9 月 14 日晚上 7 時，在安寧殯儀館舉辦。

There will be a memorial service in honor of Mr. Baba tomorrow evening.

明天晚上將舉行馬場先生的追悼會。

The visitation will be from 6-8 p.m. **at** the funeral home.

弔祭將從晚上 6 點至 8 點在殯儀館舉行。

Flowers and offerings to the family may be sent to the funeral home.

給遺屬的鮮花與祭奠品等，請直接送至殯儀館。

Gifts to the bereaved may be forwarded to the family.

給遺屬的禮品請轉交家族成員。

國家圖書館出版品預行編目資料

5句話搞定商務Email！/ Shawn M. Clankie, 小林敏彦作；
陳亦苓譯. -- 初版. -- 台北市：貝塔出版：智勝文化發行，
2011.10
　　面；公分
ISBN 978-957-729-861-4（平裝）
1.商業書信　2.商業英文　3.商業應用文　4.電子郵件
493.6　　　　　　　　　　　　　　　100018330

5句話搞定商務Email！

作　　者 / Shawn M. Clankie、小林敏彦
譯　　者 / 陳亦苓
總 編 審 / 王復國
執行編輯 / 游玉旻

出　　版 / 貝塔出版有限公司
地　　址 / 100 台北市中正區館前路 12 號 11 樓
電　　話 / (02) 2314-2525
傳　　真 / (02) 2312-3535
郵　　撥 / 19493777 貝塔出版有限公司
客服專線 / (02) 2314-3535
客服信箱 / btservice@betamedia.com.tw

總 經 銷 / 時報文化出版企業股份有限公司
地　　址 / 桃園市龜山區萬壽路二段351號
電　　話 / (02) 2306-6842

出版日期 / 2015 年 2 月初版二刷
定　　價 / 320 元
Ｉ Ｓ Ｂ Ｎ / 978-957-729-861-4

BUSINESS PERSON NO EIBUN MAIL
© Shawn M. Clankie 2010 © TOSHIHIKO KOBAYASHI 2010
Originally published in Japan in 2010 by GOKEN CO., LTD.
Chinese translation rights arranged through TOHAN CORPORATION,
TOKYO., and Keio Cultural Enterprise Co., Ltd.

喚醒你的英文語感！

請對折後釘好，直接寄回即可！

100 台北市中正區館前路12號11樓

 貝塔語言出版 收
Beta Multimedia Publishing

寄件者住址 □□□

貝塔語言出版
Beta Multimedia Publishing

讀者服務專線 (02) 2314-3535 讀者服務傳真 (02) 2312-3535
客戶服務信箱 btservice@betamedia.com.tw
www.betamedia.com.tw

謝謝您購買本書！！

貝塔語言擁有最優良之英文學習書籍，為提供您最佳的英語學習資訊，您填妥此表後寄回（免貼郵票），將可不定期免費收到本公司最新發行之書訊及活動訊息！

姓名：＿＿＿＿＿＿＿＿＿　性別：□男 □女　生日：＿＿年＿＿月＿＿日

電話：（公）＿＿＿＿＿＿＿（宅）＿＿＿＿＿＿＿（手機）＿＿＿＿＿＿＿

電子信箱：＿＿＿＿＿＿＿＿＿＿＿＿＿＿＿＿＿＿＿＿＿＿

學歷：□高中職含以下　□專科　□大學　□研究所含以上

職業：□金融　□服務　□傳播　□製造　□資訊　□軍公教　□出版
　　　□自由　□教育　□學生　□其他

職級：□企業負責人　□高階主管　□中階主管　□職員　□專業人士

1. 您購買的書籍是？＿＿＿＿＿＿＿＿＿＿＿＿＿＿＿＿＿＿＿＿

2. 您從何處得知本產品？（可複選）
　　□書店 □網路 □書展 □校園活動 □廣告信函 □他人推薦 □新聞報導 □其他＿＿＿

3. 您覺得本產品價格：
　　□偏高 □合理 □偏低

4. 請問目前您每週花了多少時間學英語？
　　□不到十分鐘 □十分鐘以上，但不到半小時 □半小時以上，但不到一小時
　　□一小時以上，但不到兩小時 □兩個小時以上 □不一定

5. 通常在選擇語言學習書時，哪些因素是您會考慮的？
　　□封面 □內容、實用性 □品牌 □媒體、朋友推薦 □價格 □其他＿＿＿

6. 市面上您最需要的語言書種類為？
　　□聽力 □閱讀 □文法 □口說 □寫作 □其他＿＿＿

7. 通常您會透過何種方式選購語言學習書籍？
　　□書店門市 □網路書店 □郵購 □直接找出版社 □學校或公司團購 □其他＿＿＿

8. 給我們的建議：＿＿＿＿＿＿＿＿＿＿＿＿＿＿＿＿＿＿
＿＿＿＿＿＿＿＿＿＿＿＿＿＿＿＿＿＿＿＿＿＿＿＿＿＿
＿＿＿＿＿＿＿＿＿＿＿＿＿＿＿＿＿＿＿＿＿＿＿＿＿＿

喚醒你的英文語感！

Get a Feel for English !

喚醒你的英文語感！

Get a Feel for English !